BEST PRACTICES FOR EDUCATION PROFESSIONALS

BEST PRACTICES FOR EDUCATION PROFESSIONALS

Volume 2

Edited by
Heidi L. Schnackenberg, PhD
Beverly A. Burnell, PhD

Apple Academic Press Inc. | Apple Academic Press Inc.
3333 Mistwell Crescent | 9 Spinnaker Way
Oakville, ON L6L 0A2 | Waretown, NJ 08758
Canada | USA

©2017 by Apple Academic Press, Inc.

First issued in paperback 2021

Exclusive worldwide distribution by CRC Press, a member of Taylor & Francis Group
No claim to original U.S. Government works

ISBN 13: 978-1-77463-700-5 (pbk)
ISBN 13: 978-1-77188-412-9 (hbk)

Library and Archives Canada Cataloguing in Publication

Best practices for education professionals / edited by Heidi L. Schnackenberg, Beverly A. Burnell.

Volume 2 published in Oakville, Ontario, Canada.
Includes bibliographical references and indexes.
Issued in print and electronic formats.
ISBN 978-1-77188-412-9 (v. 2 : hardcover).--ISBN 978-1-77188-413-6 (v. 2 : pdf)

1. Education--Methodology. 2. Educational innovations. 3. Educational psychology. I. Schnackenberg, Heidi L., editor II. Burnell, Beverly A., editor III. Title: Education professionals.
LB1027 B488 2013   370.1   C2011-908707-3   C2016-905272-9

Library of Congress Control Number: 2012935661

CIP data on file with US Library of Congress

Apple Academic Press also publishes its books in a variety of electronic formats. Some content that appears in print may not be available in electronic format. For information about Apple Academic Press products, visit our website at **www.appleacademicpress.com** and the CRC Press website at **www.crc-press.com**

CONTENTS

LIST OF CONTRIBUTORS

Aline R. Bobys
State University of New York at Plattsburgh, 101 Broad St, Plattsburgh, NY 12901, United States

Rebecca Boushie
Head Start Classroom Manager, Columbia Opportunities Incorporated, 540 Columbia St, Hudson, NY 12534, United States

Beverly A. Burnell
State University of New York at Plattsburgh, 101 Broad St, Plattsburgh, NY 12901, United States

Bradley Countermine
Beekmantown Middle School, 37 Eagle Way, West Chazy, NY 12992, United States

Emily A. Daniels
State University of New York at Plattsburgh, 101 Broad St, Plattsburgh, NY 12901, United States

John Delport
Coastal Carolina University, 100 Chanticleer Dr E, Conway, SC 29528, United States

Nanci E. Howard
Coastal Carolina University, 100 Chanticleer Dr E, Conway, SC 29528, United States

David Iasevoli
State University of New York at Plattsburgh, 101 Broad St, Plattsburgh, NY 12901, United States

Talya Kemper
California State University, 401 Golden Shore, Long Beach, CA 90802, United States

Jamia Richmond
Coastal Carolina University, 100 Chanticleer Dr E, Conway, SC 29528, United States

Heidi L. Schnackenberg
State University of New York at Plattsburgh, 101 Broad St, Plattsburgh, NY 12901, United States

Maureen E. Squires
State University of New York at Plattsburgh, 101 Broad St, Plattsburgh, NY 12901, United States

Jon Storslee
Paradise Valley Community College, 8401 N 32nd St, Phoenix, AZ 85032, United States

Edwin S. Vega
State University of New York at Plattsburgh, 101 Broad St, Plattsburgh, NY 12901, United States

Yong Yu
State University of New York at Plattsburgh, 101 Broad St, Plattsburgh, NY 12901, United States

Kerri Zappala-Piemme
State University of New York at Plattsburgh, 101 Broad St, Plattsburgh, NY 12901, United States

LIST OF ABBREVIATIONS

AA	acrylic acid
ACGIH	American Conference of Governmental Industrial Hygienists
ABA	applied behavior analysis
ASD	autism spectrum disorder
AACTE	American Association of Colleges of Teacher Education
CLT	cognitive load theory
CSS	cascading style sheet
CCBC	Cooperative Children's Book Center
CAEP	Council for the Accreditation of Educator Preparation
CBO	community-based organizations
CAEP	Council for Accreditation of Educator Preparation
CCU	Coastal Carolina University
CWI	Community, Work & Independence
CRT	critical race theory
CRC	Critical Race Curriculum
ED	ERIC database
FOP	families of promise
GSR	Galvanic skin response
GHPN	Greater Hudson Promise Neighborhood
GPA	grade point average
HCSD	Hudson City School District
IP	internet protocol
HCZ	Harlem Children's Zone
IEP	Individual Education Plans
JESPAR	*Journal of Education for Students Placed at Risk*
LOMAs	Learning Outcomes Metrics Analysis
NCATE	National Council for Accreditation of Teacher Education
SUNY	State University of New York

SEA WASP   Scientific Environmental Activities Watershed
                Alliance Sound Partnerships
SAME   social-action multicultural education
TRSE   *Theory and Research in Social Education*
WWAMH   Warren-Washington Association for Mental Health

ABOUT THE EDITORS

Heidi L. Schnackenberg, PhD, is a Professor in Teacher Education at SUNY Plattsburgh in Plattsburgh, NY. Specializing in Educational Technology, she currently teaches both undergraduate and graduate classes on the use of technology to enhance teaching and learning in the P-12 classroom, social issues in education, and ethical issues in educational technology. Her various research interests include the integration of technology into pedagogical practices and the legal and ethical implications of Western technologies in non-Western and third-world cultures.

Beverly A. Burnell, PhD, is an Associate Professor in Counselor Education at SUNY Plattsburgh in Plattsburgh, NY, and is the Employee Assistance Coordinator for the campus. She teaches graduate classes for pre-professional school counselors, mental health counselors, and student affairs professionals in college settings. She teaches courses in counseling skills, cultural contexts of counseling, career development, gender issues in counseling, and ethics in counseling practice. Her scholarly interests include academic ethics in counselor education, career development, and accreditation of professional preparation programs.

INTRODUCTION

As we sit here thinking about the idea of *best practice*, we know that concept means something different to everyone. As a parent, to Heidi, frankly, best practice in her son's school means anything that keeps him engaged, helps him learn, and that he enjoys. She doesn't really care what the "it" is, as long as he learns stuff and has fun doing it. As former teachers, our concept of best practice isn't actually all that dissimilar. We knew the curriculum we had to teach, the kids to whom we had to teach it, and it was our job to make the concepts engaging and developmentally appropriate. Simple right? Well no, not so much. While we enjoyed thinking of ways to get the students to have fun while learning, each child is different and so are their learning needs. What worked for one kid didn't necessarily work for another. And even if it worked for the kids (or some of them anyhow), our techniques might not have worked for other teachers, the administration, or even the parents. So the idea of best practice is indeed a loaded concept. But ooohhhh does it sound good! Now that we're professors residing in the ivory tower, the phrase *best practice* is a great one to trot out at academic receptions and dazzle our colleagues. It's also fun to pull out at cocktail parties if we want to alienate people and sit by ourselves. (Wow! Did you see how the Giants played last night?! No, but I just read this great book on best practices in teacher education! .... Hhhhhhmmmmm .... not exactly a crowd pleaser ...) We guess the point, though, is that the idea of *best practice* sounds good, and it is simple enough for the general public to guess what its meaning, but what exactly does it mean?

In simplest terms, *best practice* in the context of teaching and education professionals, encompasses the content that needs to be taught and the most effective pedagogies for doing so. In the third edition of their book, *Best Practice: Today's Standards for Teaching and Learning in America's Schools*, Zemelman, Daniels, and Hyde (2005) discuss how they borrow the phrase *best practice* from the medical, legal, and architectural fields. The spirit of the definition remains the same—best practice is effective, cutting-edge work in a field of study and practitioners of best practices are up-to-date on the latest research and findings in their specialty and offer their clients/patients/etc., the benefits of this current knowledge and skill set. While this definition clearly applies to the field of education in the broadest sense, Zemelman, Daniels, and Hyde (2005) go on to delineate best teaching practices in all of the various academic content areas taught in schools, including math, literacy, social studies, science, and the arts. So not only are best practices broad in scope but they are also nuanced in many subareas of study too.

So we've defined the idea of *best practices*, and even *best practice in education*, but now what? We have a concept and a neat phrase/party trick (or not, remember that cocktail party conversation killer…), but what happens with it? How do the ideas associated with *best practice* make their way from the ivory tower into classrooms and schools? Unfortunately, many educational concepts, theories, and practices created and discussed in the academy do not often find their way into the bags of tricks of the practitioners. It appears that college- and university-created knowledge may only be one directional and the elevator on that knowledge base doesn't have a variety of buttons. It heads straight to dusty college libraries (or dusty Google Scholar databases) and languishes there for egg-heads like us to write about. But it doesn't have to….

This book was created as a way for educators and education professionals to share their bodies of knowledge and the techniques and skills that work best in their particular fields. The grades and age levels that are targeted in each chapter vary, as do the types of *best practices* that are described. While this second volume of *Best Practices for Education Professionals* could be characterized as a smorgasbord of professional practice (much like the first volume), where it excels and extends the first book is in the types of practices that are described. From grassroots descriptions of the use of humor to teach dry subject matter like computer science (Storslee and Schnackenberg), to unique community placements for future special educators (Delport, Richmond, and Howard), to a current review of Chinese children's literature (Yu), the chapters included here offer unique perspectives on both content and the pedagogies with which it is taught. Undoubtedly, the most special chapter is the first, written by our dear friend and one of the most incredibly talented teachers we've ever encountered, Aline Bobys. In her contribution, she offers her reflections on best practices through a personal retrospective on her rich, 40-plus year career. We are honored to include her thoughts and stories in this book. Finally, we hope that you find the contributions included here instructive, eye-opening, and interesting, and enjoy reading it as much as we enjoyed editing it.

REFERENCES

Zemelman, S., Daniels, H., & Hyde, A. (2005). *Best Practice: Today's Standards for Teaching and Learning in America's Schools*. Portsmouth, NH: Heinemann Publishing.

CHAPTER 1

WHO KNEW?
BEST PRACTICES: A RETROSPECTIVE

ALINE R. BOBYS

State University of New York at Plattsburgh, 101 Broad St, Plattsburgh, NY 12901, United States

ABSTRACT

In a retrospective, one looks back on past events or situations, deconstructs those experiences, and passes learnings from them on to others. In this retrospective, looking back, through the lens of teaching stories, helps clarify for the author her perceptions and understandings of best practices for education professionals. It is the hope that lessons learned from these stories can help others examine and clarify their own thinking around the concept of "best practice."

THE STORIES

I am a literacy educator, a storyteller, if you will, so telling stories to communicate—to make connections—is what I do. Everything in my life revolves somehow around literacy and, as I think about writing a retrospective of my teaching practices over the course of 42 years, telling stories seems the most natural way to help me make sense of my own definition of best practices in education and to pass that information on to the next generation of educators. As education professionals—educators, counselors, speech/language pathologists, school psychologists, family and health services providers—we all have stories about the people who become a part of our lives; our students, our patients, our clients, our case families ... our common thread is the care and concern of others, and we will always remember the stories that touch our lives ... and ... with each story, the lessons we learn from those we serve. These snapshots in time become the fabric of who we are as teachers, learners, advocates, and allies.

I'm not sure where to begin, so I guess I'll start at the beginning. Let me take you back to the first years after I graduated from college, where much of my foundational learning about best practices took place, and tell you some of my stories from that time, the life lessons learned from them, and their relationship to best practices.

If you had asked me, as a recent graduate in the 1970s, about best practices for education professionals, you probably would have gotten a blank stare. The whole idea would have seemed both elusive and frustrating; what exactly is a best practice? Who gets to decide?

I always knew I wanted to be a teacher—a good teacher—but what did that mean? Learning was easy for me. I loved school. I was privileged; I had sets and sets of books to read. With each story or book I read, I learned even more about the world around me. I loved being around children. I wanted to teach them so they could love learning as much as I did. I assumed all children wanted what I wanted. I was idealistic about teaching and I was going to change the world. However, the first time I was ever in a classroom, other than as a student myself, was my senior year of my teacher education program. What did I really know about teaching? As I think about it, anything I knew about teaching was either theoretical, from textbooks and my classes, or from the heart, from my own love of learning. Who knew best practices were going to emerge from that?

And so it begins ...

Once upon a time ... long, long ago ... in a faraway place called Virginia ...

TERRIFIED BUT TAKING RISKS

I remember the day I arrived in the superintendent's office in Charlottesville, Virginia, for my first job interview for a first-grade position. The year was 1973 (yes, decades ago, as the language of this conversation will bear out), and teaching jobs were hard to come by. When I walked into the office, both the superintendent of schools and the principal of Red Hill School were there. The superintendent folded his arms, looked me up and down (yes, really), threw several folders down on the desk, and said, in a booming, burly voice: "Before me, little lady, I have the applications of a man, a teacher with six years of experience, and a teacher with a completed master's degree. Why should I hire you?"

> I was 22 years old and had no teaching experience. I was terrified. How do you answer a question like that?
> "Because I'm the best teacher you'll ever hire?" I asked, still terrified. He laughed. "And just what do you think makes you the best teacher I will ever hire?"
> I told him. I got the job.

Many times I have reflected on this story and its impact on me as an educator. I certainly could not have verbalized it at that time, but have since related this life lesson to some of what I now know to be true about best practices. Although I feel certain I talked about my love of learning and passing that love of learning on to my students, about creating engaging learning environments, learning centers, and everything I could think of that made sense to me about working with children, it seems to me there was something else that was verbalized in that interview about best practices that was even more important. The lesson? If you know who you are and what you can do to make a difference in the lives of others; if you know you are a caring, literate professional, a scholar, if you will—well educated, well versed in your field, and committed to lifelong learning not just for yourself but for your students; you then show confidence in yourself, and others will feel confident when they are around you. When you show children you are a confident learner, a risk taker, someone who is willing to take on a challenge even when terrified, might they see themselves in you, and respond in kind? Isn't this a best practice?

"LINE UP," CHAOS, AND CONSEQUENCES

I remember the day I walked into my first-grade classroom in Red Hill, Virginia, for the very first time, for my very first teaching job. I knew no one, and I didn't have a mentor to help guide me through this first experience in teaching. Try to visualize this setting: my classroom was in a circular, domed building. Inside, we didn't have traditional classrooms; rather, we had what were then called "open classrooms," in a *pod*, a domed building with no walls. There were six classes assigned to the primary pod, two first grade, two second grade, and two third grade, and each of us had our own "area" of the wide open room, delineated only by an exit door near each of our teacher's desks. The doors were used not only to define classroom areas but also, of course, used in case of emergencies or to take our students outside to the playground. Imagine six different locations with six different exit doors in a cavernous building, going around in a circle with no beginning point and no endpoint. Imagine, too, in the center of the pod a huge reading loft (my dream area), outfitted with bean bag chairs, old and well-worn sofas, bathtubs outfitted with pillows and blankets inside, and a rich and extensive library where students could explore the world outside of the hills of Virginia in which they lived. This reading loft was the only break in a space where you could, literally, see every teacher and child in grades 1–3. I was scared, excited, and totally unprepared for the twenty-two 6-year-olds in my charge, who literally scattered throughout the pod when I told them to "line up at the door." They made their way, running, hopping, skipping, and giggling, to any one of those six exit doors, and I even found a few of them hiding in the reading loft! Once again, I was terrified.

They didn't teach me "line up" in my teacher education classes.

More reflection. I didn't have any language, at that time, related to best practices, but I did figure out how this experience was going to help me become a better teacher. The lesson? Expect the unexpected. Anticipate outcomes as well as consequences of every action and interaction you have as an education professional. Embrace and understand chaos; it means things are happening. Be a supportive colleague, and mentor newcomers to your field. Tell your stories so others can learn from your experiences, good and bad, be thankful when no one gets hurt, laugh at yourself, and learn as much from those you've been prepared to work with as they learn from you. Aren't these best practices?

TAKING "TIME OUT" TO MAKE INFORMED DECISIONS

I clearly remember the "time out" area in my classroom. I had created "time out" out of a large appliance box. It was decorated inside with a soft and squishy chair, a desk, a lamp, and, of course, books ... lots of books ... you see, I didn't really like the idea of "time out," but it was considered the best method for handling behavioral issues at that time, so I felt I was expected to have a "time out" area in my classroom. I was uncomfortable with the concept, so I tried to make it a warm, welcoming, and comfortable place, which by the way, was not an aspect of the "time out" method ... and ... I remember, as if it was yesterday, the morning I put one of my first graders in "time out" and then promptly forgot about him until it was time to go to lunch. I still cringe thinking about it. He was in there for hours! Why didn't he ask to get out of "time out"??? My rationalization, of course, was that it was such a wonderful, warm place to be that he didn't want to leave. If I'm honest with myself, however, I have to consider that he was probably terrified to leave because his teacher didn't tell him he could do so.

He never moved outside of that box, and I never used "time out" again.

The lesson? Develop a philosophy, a theoretical orientation, if you will, that speaks to who you are as an education professional. Your values, dispositions, and beliefs should be apparent through your actions. Know why you're doing what you're doing, from your curriculum planning to meet the needs of each of your students to the care and concern you show to your students when they are not having their best day. If you don't believe in a particular practice, if it's not a part of your theory of the world, try something else, an alternative, to solve the problems you may encounter. Make informed decisions, think critically ... and don't be afraid to listen to your heart as well as to your head. Isn't this a best practice?

DAVID AS TEACHER AND THE ROAD TO ADVOCACY

I fondly remember David, the little boy in my class who had a hearing impairment. It was my job to teach him how to read. He couldn't hear the sounds I was trying to teach him. He couldn't participate in the language experiences I was providing. The noise in the pod eradicated any chance to tap into the hearing he did have. I found out he had never had his hearing tested and that he had many, many ear infections in his six years of life. He was so frustrated and nothing I tried was meeting his needs; I knew it but, more importantly, he knew it as well. I needed help. I referred him to the speech/language pathologist. I met with his parents. I met with the principal. I helped arrange funding for surgery through the Easter Seals Foundation because his parents didn't have insurance or the money to pay for surgery. I remember, so clearly, the look in his eyes when he came back to school after having tubes put in his ears. The classroom noise must have sounded like explosions to him. He kept covering his ears with his hands while looking at me, eyes opened wide, and smiling the whole time. He could hear and he was reading independently within 3 weeks.

David's smile touched my heart like nothing I have experienced, and I never forgot my time with him. The lesson? Do whatever it takes to get the job done. Know when you need an allied professional. Work collaboratively with others and collectively pool your knowledge, compassion, and resources. Think outside of the box. As education professionals, we have the tools, motivation, and passion to make a difference. Be a change agent in your profession. Advocate for your students, for all of the Davids with whom you will work. I have learned, from this experience and so many other experiences I have had, that our students know when we are their allies. Lives have been changed due to one caring individual. This life lesson is one I've carried with me throughout my teaching career and it has, without question, helped me define *best practice*.

WHAT'S *BEST* ABOUT CHALLENGING THE STATUS QUO?

Older and much wiser, after 2 years of teaching at Red Hill and having moved to Blacksburg, Virginia, I decided to go back to school for my master's degree in curriculum and instruction with a specialization in reading. Remember, literacy has always been the center of my life, so pursuing this degree meant I could pursue that goal I had established for helping all children fall in love with learning through the power of reading and writing. I secured another first-grade teaching position at Harding Elementary School while completing my master's degree at Virginia Tech. Rather than completing a thesis, our graduate program required a final project that forever impacted my teaching life as an educator, as well as the lives of my then first graders and all of my future students, younger and older, with whom I was to work for the next 35+ years. Our final project, in the literacy program, was to create a year-long language arts curriculum, month-by-month, week-by-week, day-by-day. The project had to include planning for the diverse needs of all students at varying reading and writing ability levels. What had to be included in the project were strategies, methods, and techniques we had learned over the course of our graduate program and for each curriculum segment, a research-based rationale for why we had selected the particular strategy, method, or technique that we did. I remember working on that project for an entire semester, papers strewn all across my living room. Sitting in the middle of the living room floor, surrounded by countless papers and books (remember this was pre-personal computers, tablets, or smartphones), I had moments of panic, pure joy, crying, and concern. Was I doing what I was supposed to be doing? My moment of clarity finally came when I thought about my first graders and how this project had the potential to change their perceptions of themselves as readers and writers.

Language arts materials for my first-grade classroom had been ordered by the previous teacher. I hated them. They were materials that didn't fit with my philosophy of teaching reading and writing. All that was ordered were workbooks and, with those materials, all my first graders would be expected to do is circle, underline, and cross out. How on earth was that going to teach them how to read and write? That was not going to work for me and I knew it wasn't going to work for them. I made an appointment with my principal, Gary, and brought to him my completed language arts curriculum program. Yet again, I was terrified, but I knew I had to take the risk of working against the status quo. I told Gary I wanted to use my own curriculum program with my first graders in lieu of the materials that had been ordered for my classroom. I explained my rationale for doing so and asked him to read through the curriculum. He took my program home with him and two days later gave it back to me. Yes. Use it. Now.

I was thrilled. The lesson? Knowledge, experience, and understanding of how children learn should be at the forefront of our thinking when making curricular decisions. The only way to interrupt institutionalized decision-making is to take a risk, to put yourself out there, and ask those in a position to help fight against the status quo to join you in doing so. As education professionals, how we choose to work with our students and the decisions we make in regard to their learning is a critically important process and not one to be taken lightly. It is our responsibility to be the caretakers not only of those we teach but also of our profession; we must understand the degree to which our practice impacts the lives of others. I feel certain this, too, is a best practice.

THANK YOU, LOVE TERI

When you've taught for as many years as I have, you receive letters from former students that touch your heart and re-energize you for the next students to come. As I was thinking about how I wanted to approach writing this chapter, I came across this letter sent to me from Teri. In her words, Teri reminds me of how those lessons learned about best practices, from so many years ago, have continued to emerge throughout my professional life.

16 May 2012

Dr. B.,

About five years ago, I was a freshman in Cohort 1 and it seems like so long ago yet at the same time, those memories are so fresh in my mind and fond in my heart. I remember my very first college class, wandering aimlessly around Sibley trying to find your room. I walked into EDU 120— scared, nervous, and unsure of what to expect—I had no idea what I'd be walking away with. You waved me in with a smile on your face and ever since then, you've never ceased to amaze me. I can't even put into words how much you mean to me and how much of an impact you have had on my life.

As I think back to all of our time together, my mind becomes flooded with so many significant and memorable learning experiences that I hold close to my heart. I want you to know I so appreciate the support and feedback you have always provided to me as your responses always put a smile on my face. I looked forward to reading your comments and questions, really showing me that the work I did was meaningful. You have taught me the importance of taking risks, how to become an advocate for change, and to be confident and proud of the work I do. With you in my life, I have learned and evolved in ways I never knew I could.

I am forever grateful for everything you have done for me—your passion, encouragement, guidance, and reassurance—always helping me to grow and succeed. I have learned a lot from you and I am so glad to have had the opportunity to get to know you as an educator and as a person. Our time together really meant a lot to me. You are a wonderful, intelligent, empathetic person and I can truly see how much you care in everything you do. Thank you for being there for me when I needed you and showing me how to reach my goals. You are such a positive influence in my life and I honestly don't know where I would be without you.

Love,
Teri

WHO REALLY KNOWS?

Who knew the life lessons I have described might someday morph into practices that might be considered *best practices*? That I would take these life lessons, these practices, into all of my professional arenas—from reading specialist to teacher educator, from curriculum developer to consultant. Do we really know what is *best* or is the answer to that question different for everyone, intertwined with the interactions and experiences we have with our individual students? Perhaps, as professionals, we each have a voice in that conversation; we all get to decide, we all get to know.

With every mandate in our field comes a flurry of activity to determine if we, as education professionals, are doing what we are supposed to be doing. The new normal, in our professional world, is the expectation that we know and are able to use evidence-based and/or research-based strategies; therefore, the relevance of identifying and defining the concept of *best practice* becomes important in trying to figure out if identified strategies meet specific, acceptable criteria. We use the term *best practice* to mean what? People in the academy as well as educators in the field have, for years, been frustrated with the use of this term without any particular research from which to determine what the concept really means. *Is what I'm doing considered a best practice? Why or why not? How do I know?*

Although we expect teachers to question the relevance of their individual approaches to teaching, as well as the methods determined by the profession to be accessible and appropriate, how do they know if these individual approaches are actually meeting what we think of as best practices? Does everyone share the same understanding? After teaching in P-12 schools and working for decades with both in-service and preservice teachers, I feel confident in saying that for every teacher, there is a different understanding. Although we may all begin with the premise that we engage in practices we think are best for students, how we get to those practices, how we define them as *best*, can be as varied as the teachers and education professionals who implement them.

To further support this conclusion, I conducted a mini-, quasi-, qualitative exercise after reading each of the author's chapters within this book to help determine if we could formulate a clear definition of best practice as a collective. Just a heads up—this was not very scientific, but was informative nevertheless. After all, each of the authors is an education professional writing about best practices from their theory of the world. My non-scientific study was to paste the text from each chapter into a word box to create a *Wordle*. A *Wordle* is an application "used for generating 'word clouds' from text that you provide. The clouds give greater prominence to words that appear more frequently in the source text (retrieved from Feinberg, 2014, http://www.wordle.net)." I discovered the key or prominent words from the chapters, although sometimes similar, all placed importance on different aspects of what we may consider to be elements of best practices. Zemelman, Daniels, and Hyde (2014) put forth, in one of the only resources available for teachers on the concept of best practice, 11 clustered principles that make up a paradigm for discussing best practices in America's classrooms. The framework suggests an intersection of student-centered (authentic, challenging, holistic, experiential), interactive (democratic, collaborative, sociable), and cognitive (constructivist, developmental, reflective, and expressive) strategies in best practice teaching (Zemelman, Daniels, and Hyde, 2014, pp. 8–18). As I studied the *Wordles* from our authors' writings, I discovered the prominent words across chapters were not identical (should they be?), many words matched those used in the principles described above, if not exactly then synonymously, and some words didn't fit within the described paradigm at all.

Chapter	Five most prominent words in descending order
Chapter 2	Students, new information, class, knowledge, cognitive
Chapter 3	Children, promise, families, education, community
Chapter 4	Students, allegory, literature, learning, concept
Chapter 5	Teachers, preservice, students, education, experience
Chapter 6	Interactive, whiteboards, technologies, students, learning
Chapter 7	Service, community, college, conference, faculty
Chapter 8	Social, studies, education, curriculum, racism
Chapter 9	Chinese, books, culture, authenticity, cultural
Chapter 10	Students, teachers, education, teaching, respondents

THE CHALLENGE

Even within this book on best practices, as educational professionals we have different understandings of what that phrase means, and our understanding may shift with a different set of circumstances, environment, or study. The challenge, then, is this: How do we ensure we are not negating, but valuing contributions of experienced educators (who may or may not have an operational definition of *best practice*) while, at the same time, celebrating the current thinking of our newer education professionals (who also may or may not have an operational definition of *best practice*)? How do we ensure that all professionals can grow from what we *do* know about best practices, while also ensuring that they have not only the right, but also the responsibility, to challenge and question scripted, institutionalized practices (that may or may not be *best*) in order to do what they know is right for the students with whom they work every day?

I have often wondered why I ... why we ... do what we do? After a few seconds the answer comes easily: We do what we do because it *matters*, because ... it's a pact that we have with each other, as education professionals, to make the world a better place ... because the world will care once they hear our stories.

KEYWORDS

- best practice
- education professionals
- retrospective
- strategies
- students
- teaching lessons

REFERENCES

1. Feinberg, J. (2014). *Wordle.* Retrieved from http://www.wordle.net.
2. Zemelman, S., Daniels, H., & Hyde, A. (2014). *Best practice: Bringing standards to life in America's classrooms* (4th Ed.). Portsmouth, NH: Heinemann.

CHAPTER 2

THE METHOD TO MY MADNESS: USING HUMOR AND SILLINESS TO IMPROVE LEARNING

JON STORSLEE[1]* and HEIDI L. SCHNACKENBERG[2]**

[1]*Paradise Valley Community College, 8401 N 32nd St, Phoenix, AZ 85032, United States*

[2]*State University of New York at Plattsburgh, 101 Broad St, Plattsburgh, NY 12901, United States*

*Jon wrote the strategies, humorous advice, and anecdotes; he uses first-person language because it tells a better story.

**Heidi wrote the literature review and more academic parts, so maybe the joke's on her.

ABSTRACT

In this chapter, comedic techniques, grounded in educational theory, are presented to illustrate that there is a middle ground where fun, and learning, genuinely mix to benefit students. Several strategies, including *beware of flying chocolate* and *flying laptops*, are explained, with examples, to demonstrate best practice in computer science instruction.

INTRODUCTION

As a big fan of stand-up comedy, I love the way great comedians use humor to make a point about topics in our society. I can remember watching Robin

Williams portray a powerful and off-beat educator in 1989s Dead Poets Society (Haft & Weir, 1989) and wishing that there were more teachers like him who could bring their material to life for students. I have been extremely blessed to have had at least two educators like that in my own life, a high school history teacher and a college physics professor. I can still remember their instruction and class activities, even though they occurred over 30 years ago!

As an educational technology professor who teaches computer science classes, I employ several cognitive strategies to enhance the learning environment in my classes. My best tools are humor and silliness. By creating a comfortable classroom environment, filled with laughter, students feel safe enough to take chances and learn from failure. My strategies are based in cognitive research, although they were developed originally to have fun while teaching and learning. And really, isn't that the point?

SCHEMA THEORY AND COGNITIVE LOAD THEORY

Although light and fun strategies on the outside, my use of humor and fun is deeply rooted in both schema theory and cognitive load theory (CLT). The idea of "schema" was originally proposed by the philosopher Immanuel Kant in 1781 (2013) in his *Critique of Pure Reason*. In this doctrine, Kant talks about transcendental schema as the way in which new categories of information are associated with previous sensory perceptions. This philosophical definition of schema is somewhat different from subsequent psychological characterizations, where schema is reframed as an organized group of concepts in the mind that is useful for categorizing incoming information (Piaget, 1926). Furthering this conceptualization, in the 1930s, Bartlett (1932) conducted research on how individuals recall new, unfamiliar, information over time. In his studies, he found that participants remembered new information in a way that reflected their own backgrounds, revealing that existing schema not only categorizes but also contextualizes how new information is remembered. This work is important because it reveals that memories are not solid, unalterable entities, but rather change and evolve to fit the experiences and knowledge we currently hold.

In the 1950s, Piaget categorized the types of schema as behavioral, symbolic, and operational (Piaget, 1953). Behavioral schema refers to the sequence of actions that represent experiences. Symbolic schemas are mental images that represent experiences. And operational schema is the internal activity that one does to manipulate thoughts, or aspects of thoughts. Building from this work, Schank and Abelson (1977) referred to a pattern of actions in a behavioral schema as a "script" and posited that all memory is episodic and organized into these scripts. In the 1980s, both Rumelhart (1980) and Mandler (1984) took the concept of schema to a more encompassing definition when they characterized it as the basic building block of cognition, and they hypothesized that all information is organized into categories in the mind. Later, Mandler admitted that this theory may be overly general as an understanding of how information is structured because we do not know much information is stored in schemas, and if it is stored and/or processed in similar ways (Mandler, 2014). He feels there is much more detail yet to be determined around the idea of schema as a cornerstone of cognition.

During the late 1970s and 1980s, schema theory, derived from the work described above, was established in the fields of literacy and reading comprehension. In their 1984 work, Anderson and Pearson found that prior knowledge (or existing schematic structures) tremendously enhances reader's comprehension of new information. They further offered that schema is constructed of subschema, called "nodes," created from prior knowledge. When new knowledge is encountered, it activates only certain nodes, but not all. Anderson and Pearson's (1984) explanation continues to the present as some of the most accepted ideas and processes involving the concept of schema.

Current research investigates electronic data models derived from schema theory (Zhuge & Sun, 2010), second-language learning and the use of schemata (Nassaji, 2007), and a reframing of schema theory in literacy instruction in conjunction with current cultural contexts and shifts (McVee, Dunsmore, & Gavelek, 2005). Mandler (2014) also updated one of his previous critical works on schema theory to reposition it as a "schema framework" since he feels more knowledge is necessary to establish it as a theory. Presently, Mandler (2014) is organizing schema into "stories" (accounts of events or people, or both), "scripts" (patterns of behavior), and "scenes" (places, backgrounds, or backdrops) to add more fine-tuned understanding to the variety of ways in which schemas work. Although a historically long-standing concept and theory that has taken a variety of avenues in the social science research, schema theory is one that continues to evolve and attempts to provide grounding as an understanding of cognition.

A related, yet newer philosophy than schema theory, CLT hypothesizes that the human brain is like a computer with a limited amount of memory available to process or encode new information (Sweller, 2010). In its most basic form, CLT refers to mental effort, or how much work it takes the human mind to do a given task. The theory categorizes this mental effort, or cognitive load, into three types: intrinsic, extraneous, and germane (van Merrienboer & Sweller, 2005). Intrinsic cognitive load is the difficulty connected to learning/understanding a specific topic (Leahy & Sweller, 2008). Extraneous cognitive load is the way in which information is introduced to a learner (Mayer & Moreno, 2010). And germane cognitive load refers to the creation of permanent knowledge, called "schemas" (Paas & Van Gog, 2006). Perhaps more simply put, extraneous cognitive load is the aspect over which teachers have control, intrinsic cognitive load is what the learner brings to the process, and potentially, germane cognitive load is what happens when the other two meet. In general, the heavier the cognitive load, the more difficult it is to comprehend information or complete a task (Clark, Nguyen, & Sweller, 2011). Therefore, it makes sense that the instructional design of materials or presentation of topics or processes work to keep cognitive load to a manageable amount. Humorous anecdotes or activities can often allow learners to encode information in a way that does not seem stressful or difficult, thus reducing cognitive load.

The techniques and strategies presented in this chapter incorporate several ways to put learners at their ease so that cognitive load is lessened and learning occurs more easily. We present a variety of these ideas here.

PUTTING THEORY INTO PRACTICE: THE METHOD AND THE MADNESS

My goal during teaching is to create multiple links or bridges from prior knowledge to new information. Because explaining the meaning of all the terminologies and syntaxes used in web design can be very dry and confusing, I use a variety of analogies to help students compare something that they do understand, to the new information that I am trying to teach to them. For instance, when I give instruction about web design, I explain to the students that when a house is being built, the walls and roof need to be constructed in the proper order, or else the house will fall down. Similarly, in web design, specific elements need to be placed in the proper order or else the webpage won't work (or be "valid"). Most students know what a house looks like, so I am building a scaffold from the prior knowledge they already possess to the new knowledge of web design structure. To continue the lesson, I then add that paint makes a house look attractive. Similarly, Cascading Style Sheet (CSS) code is the paint for the web design to make a website look pleasing to the user. The references to the paint build upon the new knowledge learning that I facilitated earlier with the house analogy.

Another example of building upon prior knowledge that I use is when I explain what an Internet protocol (IP) address does for a computer. I tell my students that the IP address is the phone number for the computer so it can send and receive calls from different computers, just like what we do when making a call to our friends and family.

The examples described above illustrate a process called scaffolding (Artino Jr., 2008). Scaffolding lowers the cognitive processing load by giving the student a starting point (prior knowledge) from which to build upon instead of forcing the student to start from scratch. I build scaffolds by relating the new information to content the students may have attained in their everyday life, or from previous classes. If the student doesn't have that same reference point, I try another analogy. The more the bridges built between prior knowledge and new material, the higher the probability of the students being able to process the content into new knowledge (Artino Jr., 2008).

One example of what I describe above is my use of jokes or silly analogies about course content as stress releasers because adventuring into new areas can be stressful. By using humor around the class material, I help build a bridge to the students' prior knowledge while stimulating multiple schemas and subschemas. This then results in a higher probability of retaining the new knowledge (Willis, 2007). I also use multimedia presentations to introduce new concepts, which have also been shown to stimulate learning (Neuman, 2009). It should be noted, however, that teachers have to be aware that employing multimedia too much can result in sensory overload (Grunwald & Corsbie-Massay, 2006).

The following are examples of things I have said or done in class, why I chose to do them, and their effectiveness. Like most instructors, I usually have a reason for everything I do during class, even though it may not seem so to the casual observer.

THE MAGIC OF THREE

I have found that when repeating a concept three times, something magical occurs: the probability of my students retaining the concept that I'm discussing increases immensely. I actually also inform my students that I will find ways to repeat most important concepts three times in order to facilitate understanding. If I can't find a different angle from which to present the material, then I'll repeat myself, but in a different tone of voice for greater emphasis. I repeat concepts in order to build stronger scaffolds to prior knowledge (Anghileri, 2006), thus more solidly embedding the new knowledge into the learner's existing schema.

BEWARE OF FLYING CHOCOLATE

I throw chocolate to my students in class who ask questions, make good comments, or catch my mistakes before tests or on assignments. Basically, I'm throwing chocolate all the time, for lots of reasons. My classroom is practically raining chocolate—Willy Wonka's got nothin' on me.

Truthfully, I use chocolate to relieve stress (as so many of us are known to do), which increases cognitive resources for learning (Shi, Ruiz, Taib, Choi, & Chen, 2007). I also want to encourage my students to ask questions in class, and if a piece of chocolate inspires them to do so, then I'm all for it. I do suspect that it's the upbeat energy in the learning environment, created by the airborne confections, rather than the sweets themselves, that enables more students to ask questions. By encouraging students when they question the material (or my interpretation of it), and for catching mistakes and making good observations, my students feel more comfortable to explore the content on their own and to trust their own conclusions and judgment about it. This is one step on the path to helping learners become independent thinkers who are able to self-assess with comfort and confidence.

I AM PSYCHIC

Sometimes when I am teaching, I tell my students that I am psychic and that I just had a vision! At first, my comment is usually met with disbelief, rolled eyes, and a few chuckles. Then I go on to tell them that it is true. I really have had a vision. I have seen the information we're discussing in my vision and it was on a test! At that point, most of the students understand what I am alluding to and start writing down the content we discussed, and a few students will even ask for more clarification about the material.

This tongue-in-cheek way of telling students that yes, this will be on the test, is more than just the typical instructor alert to pay attention to important information. My comments are actually designed to lessen the stress of taking notes and increase the certainty of what concepts may be more central to the learning objectives than others. I also hope to build a stronger sense of community in the class with the psychic comment, the test, and the importance of the content. Every group has its codes and shorthand language to signify that something is of importance. My silly comments are no different. Once I put my hand to my forehead and start chanting to the class that I've had a vision, all the students start looking at each other and smiling because they know what is coming.

DON'T TRUST ME, I MIGHT BE WRONG

I always start my first class with the announcement that students shouldn't trust me. I employ this statement to encourage critical thinking and to promote discourse in the classroom. My students usually laugh at the comment or display a very shocked look when they first hear it because many of them are used to simply accepting what they are taught in school, rather than questioning and deciding for themselves what they think. I also may use the comment about trust if I think the class is not participating enough or if I am covering a controversial topic because I want them to investigate this topic on their own. For example, sometimes I tell my classes that I am "mildly color blind" and my color combinations on a design might be off. I ask them to take my color combination advice with a grain of salt. Subsequently, if a student disagrees with a color choice I have made when teaching, I bring the color combination to a class vote so the class can decide what colors work best and don't have to simply accept what I have given them. (Along these same lines, I have also been known to choose bad color combinations purposely to see if students will contradict my opinion and as a class I will let them come up with a better combination).

My opening statements set the stage for me to be "The Guide on the Side" rather than the "Sage on the Stage." When the students know that I do not intend to be the fountain of all knowledge and that to a certain degree I will be learning along with them, it allows the students to take more ownership of the learning process and to engage with the material more enthusiastically (Ayres, 2010). A stronger scaffold is also built by the students when they critically think about and double check my information (Doering & Veletsianos, 2007), plus they are more likely to ask questions or ask for clarifications about the concepts and material. For instance, at times students will ask questions about the very specific functioning of an obscure software that I'm not familiar with (or as familiar with as I could/ should be). When this happens, I tell the students that we can figure it out together, but they need to take the lead and do some research to help them answer their questions themselves. I direct them to user manuals, online help sites, and online videos/tutorials that they can investigate and come up with their own answers to their questions. Then I ask the students to teach me what they have found out. This technique is an extremely effective way to get students to problem-solve and critically think about information that I genuinely don't know. Plus, they get the confidence-building experience of teaching me something new and I in turn get to learn fresh concepts. It's a win–win, all because I make clear to students that they absolutely need to double-check my information (or lack thereof) rather than accepting everything I say without question. Building student's intellectual curiosity is the prize in the Cracker Jack box of "Don't Trust Me, I May Be Wrong."

"I FEEL PRETTY"

I love to make wild or weird analogies during class that will build scaffolds to prior knowledge. My favorite wild analogy is that "XML (Extensible Markup Language) is HTML (Hypertext Markup Language) on steroids." Most people know what steroids are and I have established in earlier lessons what HTML is, so introducing XML in this way allows students to build on information that they have learned previously in the class, along with information that may already have from life outside the class. Using techniques like this one and building scaffolds helps students to remember things better and grabs their attention. And of course, we're all always in it for a few laughs at this point, which is important considering I need students to stay attentive through some fairly dense material on any given day.

Typically, I teach a lot of hardware/software/networking types of classes. In other words, material that can get pretty dry and seem a bit intimidating unless I do something about it. My favorite analogy is when I describe the difference between HTML and CSS. I tell my class HTML is used to identify the content and build the page, while CSS is used to make the web "Pretty, oh so Pretty." I then repeat the tagline (or "chorus" for anyone who recognizes one of the signature tunes from "West Side Story"), singing it and highlighting it with a little dance. (Yes, it's true, I've been known to do a little soft-shoe. I hear I'm not bad, although Dancing with the Stars hasn't been trying to recruit me or anything ....) I know—it's somewhat cringe-worthy—but rarely does a student walk out of my class mixing up the purposes of HTML and CSS. So, in this case, the old song and dance routine actually does work!

Analogies and characterizations are a great way to build scaffolds from prior knowledge to the new content (Mascolo, 2005). Wild, multi-modal, analogies help build stronger links, and students are more likely to remember them because the brain builds links from different areas. And, again, the stress release from the laughter generated by the analogies and characterizations frees up more cognitive resources to process the new information (Chen & Chang, 2009).

OOPS, I MADE A MISTAKE

I want my students to challenge themselves and feel comfortable making mistakes during class. When I make mistakes in class (either genuine or deliberately in order to see if they are paying attention), I encourage students to catch them and correct me. (This notion is inextricably intertwined with the "Don't Trust Me" idea.) When I am corrected on an error, I throw them a piece of chocolate and we proceed with fixing my mistake as a class. If a student makes a mistake during class, I praise them on making "a great mistake so everyone else could learn from their error" or chide them for "stealing my favorite mistake to make." I also remind them that "you don't learn anything by doing things correctly every time."

If a student is really struggling during class, I will go over and console them by making fun of their computer. I tell them that "I am sure it's your computer trying to trick you" and then I slap the side of the computer on the side and say "Bad Computer!" I also give the student kudos for putting up with such a mean computer.

By doing what I describe above, I model how I would like my students to engage in the learning process. If I make a deliberate mistake, and no one catches my error, I will stop class and I will tell them that I am sad because no one is following me closely enough to double-check my work. Alternately, I may ask my class if they are sure I am doing the project/ demonstration/process correctly. If I am able to reduce the cognitive load the students experience by showing them that it is okay for me to make mistakes in class, then hopefully they will feel better about making their own mistakes and more willing to try new things. The goal is for students to engage in rich, real-life, problems or tasks without experiencing cognitive "overload." By properly designing instructional experiences so that students can successfully problem-solve while learning, then more positive emotions will be associated with making errors (and correcting them). This then will strengthen the students' intellectual scaffolding and reduce cognitive load (van Merriënboer & Sluijsmans, 2009).

FLYING LESSONS

Learning a new concept at times can be very frustrating. If I notice that the class is having problems grasping a particular topic, I mention that in the past I've been frustrated learning new things too. For instance, I tell the students that, back in graduate school, when I was learning how to line code advanced statistical problems that never seemed to work, all I wanted to do was give my computer flying lessons! (I never actually did throw a computer back then because they were too big, but I did get a lot of satisfaction out of opening up a hard drive tower and blowing up the innards. They were only small explosions, but very satisfying at the time.) I then highlight my previous comment about flying computers by pretending to fling the laptop like a Frisbee. I mime the flying lesson to illicit a few chuckles (as well as gasps) and to lower the tension in the room. By encouraging laughter at a difficult concept, it allows students to engage with the material in a more relaxed manner. They can then focus on what they do know, or already know, and how this new knowledge fits in with current schema. (As an aside, I never actually have flung a laptop. Well, not in class anyhow …)

CONCLUSION

I remember the shock I felt when I first heard that Pew Research Center for the People and the Press found that students, under the age of 30, get their news from the Daily Show (Coscrove-Mather, 2004). After watching the show a couple of times, I realized that Jon Stewart was educating his audience about controversial topics through humor. While the skits and jokes Stewart uses are more extreme and graphic than those I use, they are great examples of how to get one's content or point across while entertaining one's audience and hopefully make them think. In this chapter, and in my own teaching practice, I took the concept of entertainment, or "edutainment" a step further. I grounded comedic techniques in educational theory, illustrating that there is a middle ground where fun, and learning, genuinely mix to benefit our students.

KEYWORDS

- **cognitive load theory**
- **education**
- **humor**
- **learning**
- **students**
- **teaching**

REFERENCES

1. Anderson, R.C., & Pearson, P.D. (1984). A schema-theoretic view of basic processes in reading comprehension. In P.D. Pearson (Ed.), *Handbook of reading research*. London, UK: Routledge.

2. Anghileri, J. (2006). Scaffolding practices that enhance mathematics learning. *Journal of Mathematics Teacher Education*, 9(*1*), 33–52.

3. Artino Jr, A.R. (2008). Cognitive Load Theory and the role of learner experience: An abbreviated review for educational practitioners. *AACE Journal*, 16(*4*), 425–439.

4. Ayres, W. (2010). *To teach: The journey of a teacher* (3rd ed.). New York, NY: Teachers College Press.

5. Bartlett, F.C. (1932). *Remembering: A study in experimental and social psychology.* Cambridge, UK: Cambridge University Press.

6. Chen, I-J., & Chang, C-C. (2009). Cognitive Load Theory: An empirical study of anxiety and task performance in language learning. *Electronic Journal of Research in Educational Psychology*, 7(*2*), 729–746.

7. Clark, R.C., Nguyen, F., & Sweller, J. (2011). *Efficiency in learning: Evidence-based guidelines to manage cognitive load.* Hoboken, NJ: John Wiley & Sons.

8. Coscrove-Mather, B. (2004). *CBS News: Young get news from comedy central.* Retrieved August 15, 2014, from http://www.cbsnews.com/news/young-get-news-from-comedy-central/.

9. Doering, A., & Veletsianos, G. (2007). Multi-scaffolding environment: An analysis of scaffolding and its impact on cognitive load and problem-solving ability. *Journal of Educational Computing Research*, 37(*2*), 107–129.

10. Grunwald, T., & Corsbie-Massay, C. (2006). Guidelines for cognitively efficient multimedia learning tools: Educational strategies, cognitive load, and interface design. Academic Medicine, 81(3), 213–223.

11. Haft, S., & Weir, P. (1989). Dead Poets Society [Motion Picture]. United States: Touchstone Pictures.

12. Kant, I. (2013). *Critique of pure reason.* Charleston, SC: Create Space Independent Publishing Platform.

13. Leahy, W., & Sweller, J. (2008). The imagination effect increases with an increased intrinsic cognitive load. *Applied Cognitive Psychology*, 22(*2*), 273–283.

14. Mandler, J.M. (1984). *Stories, scripts, and scenes: Aspects of schema theory.* Hillsdale, NJ: Lawrence Erlbaum Associates.

15. Mandler, J.M. (2014). *Stories, scripts, and scenes: Aspects of schema theory* (2nd ed.). East Sussex, UK: Psychology Press.

16. Mascolo, M.F. (2005). Change processes in development: The concept of coactive scaffolding. *New Ideas in Psychology*, 23(*3*), 185–196.

17. Mayer, R., & Moreno, R. (2010). Techniques that reduce extraneous cognitive load and manage intrinsic cognitive load during multimedia learning. In J. Plass, R. Moreno, & R. Brünken (Eds.), *Cognitive load theory*, pp. 131–152. New York, NY: Cambridge University Press.

18. McVee, M.B., Dunsmore, K., & Gravelek, J.R. (2005). Schema Theory revisited. *Review of Educational Research*, 75(*4*), 531–566.

19. Nassaji, H. (2007). Schema theory and knowledge-based processes in second language reading comprehension: A need for alternative perspectives. *Language Learning*, 57(1), 79–113.

20. Neuman, S.B. (2009). The case for multimedia presentations in learning. In Bus, A.G. & Neuman, S.B. (Eds) Multimedia and literacy development: Improving achievement for young learners, London, UK: Routledge.

21. Paas, F., & Van Gog, T. (2006). Optimising worked example instruction: Different ways to increase germane cognitive load. *Learning and Instruction*, 16(*2*), 87–91.
22. Piaget, J. (1926). *The Language and Thought of the Child*. London, UK: Routledge.
23. Piaget, J. (1953). *The Origin of Intelligence in the Child*. London, UK: Routledge.
24. Rumelhart, D.E. (1980). Schemata: The building blocks of cognition. In R.J. Spiro, B.C. Bruce, & W.F. Brewer (Eds.), *Theoretical issues in reading comprehension: Perspectives from cognitive psychology, linguistics, artificial intelligence, and education*. Hillsdale, NJ: Lawrence Erlbaum Associates.
25. Schank, R.C., & Abelson, R.P. (1977). *Scripts, plans, goals and understanding*. Hillsdale, NJ: Lawrence Erlbaum Associates.
26. Shi, Y., Ruiz, N., Taib, R., Choi, E., & Chen, F. (2007). Galvanic skin response (GSR) as an index of cognitive load. Extended abstracts of the Annual Meeting of *Computer Human Interaction*, 2651–2656.
27. Sweller, J. (2010). Cognitive load theory: Recent theoretical advances. In Plass, J.L., Moreno, R., & Brunken, R. (Eds.), *Cognitive Load Theory*. New York, NY: Cambridge University Press.
28. van Merriënboer, J.J., & Sluijsmans, D.M. (2009). Toward a synthesis of cognitive load theory, four-component instructional design, and self-directed learning. *Educational Psychology Review*, 21(*1*), 55–66.
29. van Merrienboer, J.J.G., & Sweller, J. (2005). Cognitive load theory and complex learning: Recent developments and future directions. *Educational Psychology Review*, 17(*2*), 147–177.
30. Willis, J. (2007). Review of research: Brain-based teaching strategies for improving students' memory, learning, and test-taking success. *Childhood Education*, 83(*5*), 310–315.
31. Zhuge, H., & Sun, Y. (2010). The schema theory for semantic link network. *Future Generation Computer Systems*, 26(*3*), 408–420.

CHAPTER 3

THE GREATER HUDSON PROMISE NEIGHBORHOOD: ONE COMMUNITY'S ATTEMPT TO CARE FOR ITS YOUNGEST MEMBERS

REBECCA BOUSHIE

Head Start Classroom Manager, Columbia Opportunities Incorporated, 540 Columbia St, Hudson, NY 12534, United States

ABSTRACT

People have the capacity to learn throughout their entire lives; however, poverty can have a profound effect on one's ability to learn, especially during the first five years of life. There are many factors associated with low socioeconomic status that affect a child's ability to learn and develop age-appropriate skills, including community engagement and family enrichment. During the Obama Administration, there has been an increased focus on improving the quality of life for both children and adults in poverty. One of the programs originally designed to meet this need is the Promise Neighborhood Initiative (later rebranded as the Promise Zone Initiative). This chapter outlines one community's progress from an initial, federally funded, Promise Neighborhood, to a grant-funded only Promise Neighborhood supported by the Hudson, New York community. While the goal and beginnings of the Promise Neighborhood Initiative were encouraging, the loss of funding for communities that were early adopters can be devastating. Described here is how one, small, urban community handled this abandoned federal initiative.

INTRODUCTION

The effect of poverty on children has many implications for their future academic success. The notion that there are vast differences in the education that children in the United States receive is not a new idea. There have been many studies that demonstrate the inequalities that exist in public education and the resources that are allocated to children through the school systems (Carter and Welner, 2013; Condron and Roscigno, 2003; Logan et al., 2012; Partanen, 2011). The quality of a child's education is deeply rooted in their geographic location and their parent's socioeconomic status. In his book *Savage Inequalities*, Jonathan Kozel (1991) examines the vast disparities that exist within public schools that are literally only several geographic blocks from one another. The differences occur in all aspects of the school—the physical building itself, the educational materials that are available, the technological resources that are in place, the size of the classes, and the quality of the teachers. Most of the research in the past few decades have been focused on the inequalities of the K-12 public school system and the implications of those inequalities on the academic success of children.

Recently, early childhood education has become a hot topic in our nation, with many politicians and policy-makers weighing in on the issue. Stemming from these debates is the question of the effect of poverty on children from birth through age 5. How does poverty effect how children are able to succeed in "formal schooling?" In President Obama's 2014 State of the Union address he called for more spending at the federal level so that all children have access to high-quality early childhood education. What he proposed was the Preschool for All Initiative which would partner with all states to serve children from low- and moderate-income families. It would allow for more access to high-quality early childhood education to those children and families (Early Learning/The White House, 2015). Many states have picked up on this issue as well and have begun to increase the funding they allocate in their budgets for early childhood education. According to the Education Commission of the States, 30 of the 40 states that support early childhood programs have increased their level of funding in the 2013–2014 fiscal year. While putting more money into early childhood education is certainly important, the way(s) in which that money is best used is still in question.

Perhaps the first step in deciding how an increase in early childhood education funding should be spent is to examine the root of why early childhood education is important in the first place. What exactly is it that is creating the gap between socioeconomic classes and in a child's ability to enter kindergarten ready to academically, emotionally, and socially succeed? Answering this question may be easier said than done as research has pointed out that a child's ability to develop can be attributed to both socioeconomic status as well as other variables, and usually these variables are not independent but rather intertwined with each other (Hoff, 2003). A review of the literature will shed some light on what is known so far about factors that contribute (or not) to successful early learning experiences for children.

LITERATURE REVIEW

Over the past 50 years, children's academic achievement gap has been steadily widening and deepening. In 1966, a study done by James Coleman, known as the Coleman Report (Coleman, 1966), was one of the first to show the correlation between a family's social economic status and its children's academic achievements (Reardon, 2011). The income gap between upper and lower income families has also increased in the past 50 years, and with it, the achievement gap between children from these respective families has also grown. Reardon (2011) points out that "the achievement gap between children from high and low income families is roughly 30 to 40 percent larger among children born in 2011 than among those born twenty-five years earlier." It now is the most important aspect for determining future academic success.

From a biological perspective, there is some evidence that a family's socioeconomic status has an effect on a child's academic achievement. Harvard's Center on the Developing Child has undertaken a study into some of the physiological concepts of children's development and how to use those concepts to create a better system for early learning. The most important finding is that "brains are built over time" but a large portion is built from the time of birth to age 5 (National Scientific Council on the Developing Child, 2007). There needs to be interactions, what they have termed as "serve and return," between children and caring adults in order to properly hardwire an individual's brain. Once the wiring has been set, it becomes increasingly difficult to rewire. Thus, the quality of early childhood education becomes extremely important.

The Harvard Center finding could also be construed to be a call for more parent education and their exposure to best practice, which needs to begin at a child's birth, if not before (National Scientific Council on the Developing Child, 2007). A mother's education level is especially important and studies show that there is a high correlation between a mother's education and their children's cognitive development. Even for children who have other risk factors, being born to a well-educated mother will help increase their cognitive development (Walker et al., 2011). Educating parents could help to break the cycle of poverty and subsequently shrink the student achievement gap. Children born into poverty typically results in them having less of a likelihood of going farther with their education. Less education usually means that when these children are adults, they attain jobs that will garner them less income than a peer with a higher education and a professional job. Lower paying jobs then push these children, as adults, back into a life of poverty for their own children (Walker et al., 2011).

A well-known study done by Betty Hart and Todd Risley (2003), the 30 Million Word Gap, illustrates the importance of the vocabulary that a family uses and the resultant learning gains attained by their children. Hart and Risley observed 42 families with children who were learning to talk (1–2 year olds) from varying socioeconomic status, for 2.5 years. They observed the families in their homes for 1 hour each month. They separated the participants of their study into three broad categories—professional families (higher socioeconomic status), working-class families (middle/lower socioeconomic status), and families on welfare. Their major finding was that children from the welfare families had smaller vocabularies and were not acquiring words at the same pace as those children from higher socioeconomic backgrounds. The 30 million word gap comes from an estimation that, over the course of 4 years, a child from a professional family would have heard approximately 45 million words, a child from a working class family approximately 26 million words, and a child from a welfare family 13 million words.

An interesting result of the Hart and Risley study comes in the form of the type of language heard by the children. Professional families were much more likely to give their children affirmatives over the course of the observation—there was on average a 6:1 ratio of affirmatives to discouragements. This ratio dropped significantly in the working-class family to a ratio of 2:1 and went conversely in welfare family—children heard two *discouragements* for every one *affirmation* (Hart and Risley, 2003). Although the researchers are careful to point out that each family "nurtured their children and played and talked with them. They all disciplined their children and taught them good manners and how to dress and toilet themselves. They provided their children with much the same toys and talked to them about much the same things" (Hart and Risley, 2003). Therefore, then the assumption is to be made that it was the quality of the interactions that was creating the differences. A child from a higher income family have parents, mothers in particular, who "speak in longer utterances, using richer vocabulary, and producing more complex sentences than lower SES mothers" (Pungello et al., 2009). Parents are children's first, most important and probably most formative teachers, and as this study demonstrates children tend to speak like their parents.

Another contributing factor to the widening achievement gap is the availability of educational materials, specifically books, in the homes of children under the age of 5. Some research estimates that "nearly two thirds of low-income families in the United States own no books" (Reading Is Fundamental, 2015). One of the barriers to families owning books is the cost of the books themselves. A high-quality picture book has a retail value of approximately $15–$20 (Bornstein, 2011). When having to make choices on an already stretched budget, the picture book is not usually seen as a necessity. Unfortunately, this same lack of books is also sadly noticeable in low-income preschool programs. The lack of funding in these schools forces school personnel to make choices as to what is necessary and what isn't, and similar to low-income homes, books usually do not fall in the necessary category (Bornstein, 2011).

A study done by M.D.R Evans et al. (2010) looked at 27 different nations and compared the number of books in a home to how many years of education the children from these homes obtained. The study was done internationally to control for varying factors such as culture and political ideologies. The researchers found that in every country in their study, more books in the home meant that the children had more years of schooling. "Home Library size has a very substantial effect on educational attainment, even adjusting for parents' education, father's occupational status, and other family background characteristics. Growing up in a home with 500 books would propel a child 3.2 years further, on average, in education, than would growing up in an otherwise similar home with few or no books" (Evans et al., 2010). Having books in the home (and school) relays an importance of reading to the children, even if their parents are illiterate. According to the study, the number of books in the home transcends all other factors including that of the influence of time and technology.

Interestingly, gender also plays a role in the achievement gap. Female children from high socioeconomic families are the leading cause of the ever widening gap in the attainment of higher education (Bailey, 2011). Research suggests that children, particularly females, from families with a higher socioeconomic background are being encouraged and groomed for higher education. However, children from lower socioeconomic families are not being encouraged or prepped in the same way for higher education (Bailey, 2011). As mentioned previously, one of the reasons that is theorized about the cause of the achievement gap in education is that children in middle to upper socioeconomic families have more access to, and the parents are more likely to seek out, high-quality early childhood education as well as other socializing opportunities to increase their children's cognitive abilities and social skills (Reardon, 2011). This is not as easy of a task for a family who is faced with financial difficulties and other hardships. Quality early childhood education programs are few and far between and when they are available they are usually extremely expensive. (Speaking from personal experience—as both a working professional and a mother of two young children—for my children to attend a full-day preschool program at the same time costs our family more than the mortgage on our house).

One of the most basic ways to improve the quality of early childhood education and insure that all children are receiving the same type of support is through the professionals that work in the field. The level of education of professionals in a preschool program varies tremendously. A job search shows how the requirements for an early childhood educator vary from a high school diploma to an associate's to a bachelors, depending on the program that is hiring. There is evidence that also points to the fact that the more urban and poverty stricken an area is, the less likely that the educators in the local early learning centers are to hold any kind of advanced degree (Cunningham et al., 2009). Given the shift in focus at the elementary level, where young children are taking reading and writing-intensive standardized tests, thus placing increasing demands on what they know entering school, it is as important as ever that professionals at the early childhood level are aware and know how to effectively teach preacademic skills (Cunningham et al., 2009).

The overwhelming evidence as described above that put children in America at a disadvantage from the start has caused many professionals and politicians to begin addressing the problem. There are numerous intervention programs on all levels—local, state, and federal to address the issue of socioeconomic status and its correlation to poor academic success. While those interventions are making gains on their own, there is one intervention program whose aim is to join all current resources in a community, find the "gaps" in those resources, create solutions to those gaps, and create a "seamless pipeline of services" so that no child will fall through the cracks; this program is the Promise Neighborhood.

THE GREATER HUDSON PROMISE NEIGHBORHOOD

The National Early Literacy Panel released findings in 2004 that suggested that there are 11 factors in literacy and language acquisition that directly correlate with children's reading and overall success in future schooling (Strickland and Shanahan, 2004). Among those areas of preschool skills that were found with high correlations to future academic and reading success were oral language, alphabetic knowledge, print knowledge, and invented spelling. In response to this research, the United States Department of Education (under the Obama Administration) established the Promise Neighborhoods program (United States Department of Education, 2015). "The program is intended to significantly improve the educational and developmental outcomes of all children in our most distressed communities, including rural and tribal communities, and to transform those communities by:

1. Supporting efforts to improve child outcomes and insure that the outcomes are communicated and analyzed on an ongoing basis by leaders and members of the community;

2. Identifying and increasing the capacity of eligible entities that are focused on achieving results and building a college-going culture in the neighborhood;

3. Building a continuum of academic programs and family and community supports, from the cradle through college to career, with
   a strong school or schools at the center;
4. Integrating programs and breaking down agency 'silos' so that solutions are implemented effectively and efficiently across agencies;
5. Supporting the efforts of eligible entities, working with local gov-
   ernments, to build the infrastructure of policies, practices, systems,
   and resources needed to sustain and 'scale up' proven, effective
   solutions across the broader region by the initial neighborhood; and
6. Learning about the overall impact of Promise Neighborhoods and
   about the relationship between particular strategies in Promise
   Neighborhoods and student outcomes, including a rigorous evaluation
   of the program." (Promise Neighborhoods Institute at PolicyLink,
   2014).

According to the Department of Education website, there have been
three grant cycles offered through the Promise Neighborhood Initiative
resulting in 12 implementation grantees and 46 planning year grantees.
In 2011, the city of Hudson and its surrounding areas, located in New York
state, became one of the 15 communities in the United States awarded
a planning year grant that year (Promise Neighborhoods Institute at
PolicyLink, 2014). This particular Promise Neighborhood planning grant
encompassed the Hudson City School District, with the city of Hudson
being in the center. The Hudson City School District (HCSD) serves
a very unique blend of children from urban, suburban, and rural areas.
According to the Greater Hudson Promise Neighborhood's (GHPN) grant
application, 36% of all families in the HCSD live below the poverty line;
that number increases to 49.9% for families with children under the age
of 5 (these figures were obtained from the 2009 American Community
Survey). These numbers are significantly higher than the statewide average; which are 16.9% of all families and 20.2% of all families with children under the age of 5. Additionally, another major concern facing the
Hudson area, according to the GHPN's grant application, was that of the
availability of quality early learning opportunities. Approximately only
40% of children under the age of 5 would be able to participate in a formal
early learning program (this includes family daycares). These statistics
and the community profile they created made Hudson an excellent recipient for the Promise Neighborhood planning grant.

During 2012, the "Greater Hudson Promise Neighborhood," as this program was titled, conducted extensive research on community strengths and needs. Focus groups, community assessments, and other data collection activities occurred and a "solutions plan" was written to be submitted if and when a new grant was released. When the project first started, there were four main components that were studied and researched. They were Early Learning, Supporting Parents, Out of School Time, and School Reform. Each of these components encompassed a unique theme while also overlapping into the other groups. Members of the community that were already involved in a certain area were invited to participate in monthly "working groups." Many members of these groups participated in more than one focus area. It was during these working groups that solutions were vetted, created, and ultimately prepared. Having community participation and buy in is a central theme to the Promise Neighborhood project, as is creating a "seamless pipeline of services" from "cradle to college/career."

The focus groups were mostly made up of professionals from their respective fields; the GHPN's goal was to have everyone in the community have a voice in the direction of the program. This was accomplished by holding community dinners, conducting door-to-door surveys, and being a presence in the community. In order to increase participation in the project, especially by general community members, the GHPN held several 1-day activities. These included a community block party and a community clean-up day. Besides the community, the GHPN was involved in activities in the school district (both by participating in school initiated events and by holding GHPN events in the school).

As part of the final solutions plan, the Early Learning portion of the project consists of five main components, all of which have a literacy focus. They are Families of Promise (FOP), Parent–Child Home Program, Mobile Learning Center, a GHPN-led Early Childhood Learning Center, and a "Get Ready for Kindergarten" program. In addition, there was an effort to create a Provider Council, which was to serve as an information-sharing venue that included all of the agencies and people who served children under the age of 5. All of the programs proposed in the solutions plan for the GHPN are deeply rooted in evidence-based community and education programs that have been proven successful with young children. The programs are described below.

The nationwide Parent–Child Home Program began in 1965 and serves "under-resourced" families. Their website describes the program as "an evidence-based early literacy, parenting, and school readiness model ..." that is "committed to closing the achievement gap by providing low-income families the skills and materials they need to prepare their children for school and life success" (Parent-Child Home Program, 2015). The Parent–Child Home Program has several factors that make it appealing to communities in need like Hudson. The first is that they work with children and families from ages 16 months to 4 years old. Once a family is enrolled in the program, they receive 30-minute visits twice a week for 2 years, with the children aging out no later than the age of 4. The Parent–Child Home Program hires local professionals as staff, and these individuals are already a part of the community and families cultural background. This practice removes obstacles such as language barriers, cultural misunderstandings, and misinterpretations of customs and traditions. Another key component of the Parent–Child Home Program is that it utilizes a "modeling" technique. The literacy specialists that work with the families model ways to engage the children rather than lecture the parents and families on best practices. During each visit, a staff member will bring a book or other educational toy and use that as the basis for the visit. Together, the staff member and child/children explore how to communicate, play, and engage in developmentally appropriate activities (Parent-Child Home Program, 2015). The families then keep the book or educational toy to add to their home libraries and/or resources.

A second initiative that was proposed through GHPN was due to the lack of public transportation available in the Hudson area, including a deficit within the city of Hudson itself. Since families were not able to easily get to high-quality early learning opportunities due to lack of transportation than it seemed logical to bring it to them! Therefore, the Mobile Learning Center was proposed as a solution to this issue. There are several areas throughout the country that successfully do similar initiatives, such as Library2Go! in King's County or the more widely known "Book Mobile" idea that serves populations throughout the United States. Interestingly, the National Center for Education Statistics estimated that there are 864 Book Mobiles nationwide (Warburton, 2013). Similar to the idea of the Book Mobile, the Mobile Learning Center would bring high-quality books and other early learning and educational resources to all areas of the Hudson community. The Mobile Learning Center would be staffed by educated professionals in the field of early childhood education so that they could do outreach and education as they navigated the center to various families throughout the area.

A third proposal stemming from the GHPN solutions plan was that of a GHPN-led Early Learning Center. The center would serve children from 6 weeks old to age 5. One of the many problems facing families who are interested in having their children in early learning programs is the length of time the programs span each day. Many programs run for 5–6 hours per day, or sometimes even less. It is a hardship on families to enroll children in these programs and then find additional care for the rest of the day if parents/caregivers need to work. As part of the GHPN Early Learning Center, there would be "wrap around" services that would allow the children to participate in outside early learning opportunities, such as the Hudson City School District's UPK program (currently a 2.5 hour program) and then attend the GHPN's Early Learning site for the rest of the time that a family would need. All of the tuition and fees for the site would be based on a sliding fee scale, according to parent/caregiver/family income and need. For the children who would attend the GHPN's Early Learning site, there would be ample access to books and other learning materials, as well as staff who were a well-rounded, educated, group of teachers.

Related to the Early Learning Center is the proposed "Get Ready for Kindergarten" proposal. Many children were entering the Hudson City School District underprepared for kindergarten. Because of this, the Hudson City School District holds a Kindergarten Academy every year for 2 weeks prior to the start of the school year. This program allows children who attend to become familiar with the school, the teachers and it also has an academic component. The proposed "Get Ready for Kindergarten" program would differ from the Kindergarten Academy in both length and intensity. It was designed as an intensive, 6-week, summer program that would be filled by children who were at the most risk, as identified by their kindergarten screening results. One major obstacle to the "Get Ready for Kindergarten" program would be the ability of parents and caregivers to get their children to the location at which the program will be offered.

The final program proposed in GHPN's early learning section of the solutions plan was the idea of FOP. FOP is a 6-week program for expecting new parents. It is modeled after the very successful Baby College of the Harlem Children's Zone (Harlem Children's Zone, 2015). (The entire Harlem Children's Zone (HCZ) initiative, founded by Geoffrey Canada, is actually the basis for the Promise Neighborhood project.) During the FOP program, each participant (parents/families are very much encouraged to attend together; however, historically it has been traditionally mothers; there are other groups that are focusing on increasing a father's involvement with their child) is asked to complete a pre- and posttest that comprises questions relating to how they are going to raise their baby. Many questions refer to how often they will read to their child, the availability of books in their homes, and their intentions on using childcare or other early learning opportunities. Each week during the face-to-face program, a new topic is discussed, which ranges from children's developmental milestones to safety in the home. During several weeks, experts come in and provide deeper knowledge into their particular topic, such as the nutritional or medical needs of newborns. However, every session begins in the same way, with each participant receiving a book that is read aloud to the group to model how to read to children. The participant can then bring the book home to start and/or add to their collection of children's literature. There are also other incentives that are offered, such as transportation, childcare, breakfast and lunch, and other gifts that go along with the topic of the week (i.e., during the week on safety, each family receives items such as outlet covers and safety gates). During the first round of FOP if a parent/family had perfect attendance, they were entered into a drawing to win a free month of rent.

A unique component of the FOP program is that it is not a court-mandated parenting education course. There is a stigma that may come with being "forced" into parenting education because everyone comes from their own unique background and with their own set of skills. FOP embraces those differences to find strengths as well as educating families on what current best practices are. It also serves as a support group for people who may be lacking a support system of their own. A place where new and expecting parents can share their fears, frustrations, hopes, and triumphs for their children with others who are in a similar place. By creating a support system of this nature, it sets the foundation for these parents to bond and share their experiences with each other throughout the different stages of their children's lives.

THE GREATER HUDSON PROMISE NEIGHBORHOOD TODAY

While in the planning year, the GHPN was able to receive a 1-year extension, until December 2013, in order to complete research on the initiatives
outlined in the solutions plan. In August of 2013 along with the Mental
Health Association of Columbia and Greene Counties, GHPN applied
for and received an AmeriCorp grant. Since then there have been no further grant opportunities through the Promise Neighborhood Initiative to
receive an implementation grant. Through the hard work and dedication of
the GHPN director and other community members, the work of the GHPN
has continued in spite of this lack of funding. The project was kept afloat
from January of 2013 through March 2013 with no budget or funding
source. Staff was greatly reduced to only the director. In March of 2013,
the tides began to turn in favor of the project when it received a large grant
through the Dyson Foundation as well as smaller grants from Columbia
County, New York. The biggest boost to the GHPN's current budget came
from being awarded a Community Schools Grant in July of 2014.

Presently, the GHPN has several programs that come from the original
project's solutions plan. The Mental Health Association of Columbia and
Greene Counties is the lead fiscal agent for the project. These programs
include FOP, which as mentioned before is going into its sixth cycle of
participants. Besides FOP, GHPN also has the following initiatives up and
running the following:

- POPS (Promise Opportunity Parenting Strengths). This w is a program for fathers stemming from the current National Fatherhood
  Initiative to increase the amount of participation that a father has in
  their children's lives.
- AmeriCorp and Jr. AmeriCorp. This allows mentors to work
  throughout the community, including in the intermediate school of
  the HCSD, the GHPN office, and the Mental Health Association of
  Columbia and Greene Counties.
- HOST (Hudson's Out of School Time Collaborative). This initiative
  is a group of people who are all involved in activities that take place
  outside of the school day that are for children in the Hudson area.
  Their focus is to combine their resources and make sure that programs are available and known to all eligible children.

- Greater Hudson Initiative for Children of Incarcerated Parents.
- Two mentoring programs—one with the local Cornell Cooperative Extension that includes programs such as 4H mentors, Family night-out events, and Youth and Families with Promise. The other group of mentors is part of a larger mentoring group within Columbia County and focuses on mentoring youth with incarcerated parents.
- Community Schools Initiative. This works in conjunction with the HCSD. This project employs a full-time director and three full-time parent coordinators to work as a liaison between the families and the school district.

In addition to the initiatives described above, GHPN's main goal of being a positive presence in the community also continues. This occurs through the continuing of community projects such as the biannual community cleanup and the annual community block party. The GHPN's work has been a major push in the community to increase people's access to different programs that will help move the community forward, including those such as FOP which helps to alleviate the inequality of children's experiences under the age of 5.

THE PROMISE NEIGHBORHOOD INITIATIVE AND PROMISE ZONES

Currently, the Promise Neighborhood Initiative has neither released other funding nor offered any more grant opportunities to the GHPN. In his 2013 State of the Union address, President Obama outlined a plan to designate several at-risk communities as places that would partner with the federal government, as well as local area businesses, in an effort to create jobs, increase education, diminish crime, etc. (Whitehouse, Office of the Press Secretary, 2013). Communities must apply to become part of this initiative, and if awarded, they are designated as "Promise Zones" (U.S. Department of Housing and Urban Development, 2015). Nationwide, as of January 9, 2014, President Obama announced the initial five sites to receive "Promise Zone" designations. These sites are Philadelphia, Pennsylvania; Los Angeles, California; southeastern Kentucky; the Choctaw Nation in Oklahoma; and San Antonio, Texas (Promise Neighborhoods Institute at PolicyLink, 2014). In May of 2015, the president announced another 15 sites to become

Promise Zones. These include Camden, New Jersey; Hartford, Connecticut; Indianapolis, Indiana; Minneapolis, Minnesota; Sacramento, California; St Louis (and St Louis County), Missouri; South Carolina Low Country; and Pine Ridge Indian Reservation of the Ogala Sioux Tribe in South Dakota (U.S. Department of Housing and Urban Development, 2015).

Although a seemingly powerful initiative, Promise Neighborhoods appears to have become the predecessor to Promise Zones, and the focus has changed slightly due to the department(s) sponsoring the initiatives. Promise Neighborhood grants were awarded through the Department of Education, with, of course, a focus on educational initiatives. Promise Zones are through the Department of Housing and Urban Development and have an emphasis on changing the face of poverty-stricken neighborhoods themselves. Part of the change in the nature of the related, but yet different, initiatives could be the controversy that emerged during the beginning of Promise Neighborhoods. In 2010, the Brookings Institution published a report admonishing the Obama administration for attempting to replicate the HCZ model in Promise Neighborhoods (Whitehurst and Croft, 2010). The report contended that HCZ and its associated initiatives haven't demonstrated any improvement in student achievement. It appears the Brookings report was enough to plant the seed of doubt at the very beginning as to whether or not Promise Neighborhoods would help our nations impoverished, and their children. At this time, although information still exists on how to set up a Promise Neighborhood and a Promise Neighborhoods Institute at PolicyLink (PolicyLink is a California-based nonprofit organization) has been created to support communities wishing to become Promise Neighborhoods (Promise Neighborhood Institute at PolicyLink, 2011), no more funding appears to have been made available since 2012.

CONCLUSION

There is a cyclical nature to life. There is also a cycle to poverty (Gopal and Malek, 2015). As demonstrated through the idea of the "cradle to career" approach offered by the Promise Neighborhood Initiative and through some of the various studies cited in this paper, individuals more often than not follow in the footsteps of their parents or caregivers, through both nature and nurture. Therefore, as a nation we must focus our attention on leveling that so-called playing field so that all children are able to dream big, achieve their dreams, and in the case of the impoverished, succeed beyond what their parents have aspired to. If our current state of education including the amount and quality of our early education system continues, then we will be eliminating (or perhaps even segregating) a large portion of our population from this fundamental right to dreaming and succeeding. Sean Reardon writes "As the children of the rich do better in school, and those who do better in school are more likely to become rich, we risk producing even more unequal and economically polarized society" (2011). Investing funding, *and continuing to fund*, early childhood education and initiatives like Promise Neighborhoods, will insure the quality of these programs, and their availability to everyone, regardless of socioeconomic status. Focusing on youth and families will go a long way toward breaking the cycle of poverty.

KEYWORDS

- **early childhood**
- **early intervention**
- **education**
- **learning**
- **literacy**
- **promise neighborhoods**

REFERENCES

1. Bailey, M.J., & Dynarski, S.M. (2011). Gains and gaps: Changing inequality in U.S. college entry and completion. *National Bureau of Economic Research*. Retrieved June 6, 2015, at http://www.nber.org/papers/w17633.
2. Bornstein, D. (2011). A book in every home, and then some. *The New York Times*. Retrieved June 6, 2015, at http://opinionator.blogs..nytimes.com/2011/05/16/a-book-in-every-home-and-then-some/?partner=rss&emc=rss&_r=0.
3. Carter, P., & Welner, K.G. (2013). *Closing the opportunity gap: What America must do to give every child an even chance*. New York, NY: Oxford University Press.

4. Coleman, J.S. (1967). *The concept of equality of educational opportunity.* Baltimore, MD: Johns Hopkins University. Retrieved from ERIC database (ED012275).

5. Condron, D.J., & Roscigno, V.J. (2003). Disparities within: Unequal spending and achievement in urban school districts. *Sociology of Education,* 76, 18–36; doi: 10.2307/3090259.

6. Cunningham, A.E., Zibulsky, J., & Callahan, M.D. (2009). Starting small: Building preschool teacher knowledge that supports early literacy development. *Read Writ,* 22, 487–510; doi: 10.1007/s11145-009-9164-z.

7. Early Learning/The White House (2015). *Education: Knowledge and skills for the jobs of the future.* Retrieved June 17, 2015, at www.whitehouse.gov/issues/education/early-childhood.

8. Evans, M.D.R., Kelley, J., Sikora, J., & Treiman, D.J. (2010). Family scholarly culture and educational success: Books and schooling in 27 nations. *Research in Social Stratification and Mobility,* 28(*2*), 171–197; doi: 10.1016/j.rssm.2010.01.002.

9. Gopal, P.S., & Malek, N.M. (2015). Breaking away from the cycle of poverty: The case of Malaysian poor. *The Social Science Journal,* 52(*1*), 34–39.

10. Harlem Children's Zone (2015). *The Harlem Children's Zone.* Retrieved July 23, 2015, at http://www.hcz.org.

11. Hart, B., & Risley, T.R. (2003). The early catastrophe: The 30 million word gap by age 3. *American Educator,* 1–9. Retrieved from ERIC database (EJ672461).

12. Hoff, E. (2003). The specificity of environmental influence: Socioeconomic status affects early vocabulary development via maternal speech. *Child Development,* 74(*5*), 1368–1378.

13. Kozel, J. (1991). *Savage Inequalities: Children in America's Schools.* New York, NY: Crown Publisher.

14. Logan, J.R., Minca, E., & Adar, S. (2012). The geography of inequality: Why separate means unequal in American public schools. *Sociology of Education,* 85(*3*), 1–36; doi: 10.1177/0038040711431588.

15. National Scientific Council on the Developing Child, Center on the Developing Child, Harvard University (2007). *The Science of Early Childhood Development: Closing the Gap Between What We Know and What We Do.* Retrieved July 23, 2015, at http://developingchild.harvard.edu/index.php/download_file/-/view/67/.

16. Parent-Child Home Program (2015). *The Parent-Child Home Program.* Retrieved July 23, 2015, at http://www.parent-child.org.

17. Partanen, A. (2011). What Americans keep ignoring about Finland's school success. *The Atlantic.* Retrieved July 23, 2015, at http://www.theatlantic.com/national/print/2011/12/what-americans-keep-ignoring-about-finlands-school.

18. Promise Neighborhoods Institute at PolicyLink (2011). *What Is a Promise Neighborhood.* Retrieved June 18, 2015, at http://www.promiseneighborhoodsinstitute.org/What-is-a-Promise-Neighborhood.

19. Promise Neighborhoods Institute at PolicyLink (2014). *Promise Zones.* Retrieved June 18, 2015, at http://promiseneighborhoodinstitute.org/Promise-Neighborhoods-Movement/Promise-Zones.

20. Pungello, E.P., Iruka, I.U., Dotterer, A.M., Mills-Koonce, R., & Reznick, J.S. (2009). The effects of socioeconomic status, race, and parenting on language development in early childhood. *Developmental Psychology,* 45(*2*), 544–557; doi 10.1037/a0013917.

21. Reading Is Fundamental (2015). *Reading Is Fundamental.* Retrieved July 23, 2015 at http://www.rif.org
22. Reardon, S.F. (2011). The widening academic achievement gap between the rich and the poor: New evidence and possible explanations. In Duncan, G.J., & Murnane, R.J. (Eds), *Whither Opportunity: Rising Inequality, Schools, and Children's Life Chances* (pp. 91–115). New York, NY: Russell Sage Foundation.
23. Strickland, D.S., & Shanahan, T. (2004). Laying the groundwork for literacy. *Educational Leadership,* 61(6), 74–77.
24. U.S. Department of Education (2015). *Promise Neighborhoods.* Retrieved July 23, 2015, at http://www2.ed.gov/programs/promiseneighborhoods/index.html.
25. United States Department of Housing and Urban Development (2015). *Promise Zones.* Retrieved June 18, 2015, at http://portal.hud.gov/hudportal/HUD?src=/program_offices/comm_planning/economicdevelopment/programs/pz.
26. Walker, S.P., Wachs, T.D., Grantham-McGregor, S., Black, M.M., Nelson, C.A., Huffman, S.L., Baker-Henningham, H., Chang, S.M., Hamadani, J.D., Lozoff, B, Meeks-Gardner, J.M., Powell, C.A., Rahman, A., & Richter, L. (2011). Inequality in early childhood: Risk and protective factors for early child development. *Lancet,* 378 (9799), 1325–1338; doi: 10.1016/S0140-6736(11)60555-2.
27. Warburton, B. (2013). Delivering the library. *Library Journal.* Retrieved June 5, 2015, at http://www.lj.libraryjournal.com.
28. Whitehouse, Office of the Press Secretary (2013). *Remarks by the President in the State of the Union Address.* Retrieved June 18, 2015, at https://www.whitehouse.gov/the-press-office/2013/02/12/remarks-president-state-union-address.
29. Whitehurst, G.J. R., & Croft, M. (2010). *The Harlem Children's Zone, Promise Neighborhoods, and the Broader, Bolder Approach to Education.* The Brookings Institute. Retrieved June 18, 2015, at http://www.brookings.edu/research/reports/2010/07/20-hcz-whitehurst.

CHAPTER 4

MOVING FROM CONCRETE TO ABSTRACT UNDERSTANDINGS: STUDYING A CONCEPT'S DEVELOPMENT IN CONTEXT

MAUREEN E. SQUIRES

State University of New York at Plattsburgh, 101 Broad St, Plattsburgh, NY 12901, United States

ABSTRACT

According to Vygotsky, to *fully* understand a concept, one must understand the conditions of the concept's origin and development. I believe that, in many cases, students do not grasp conceptual understandings. Rather, in an era of fast-paced, inch-wide, mile-deep curriculum coverage, students learn shortcuts or memorize bits of information on a surface level. Such limited understandings, or misunderstandings, impede cognitive development. The purpose of this chapter is to (1) present Vygotsky's learning theory, (2) demonstrate how Vygotsky's principles can be applied in the high-school classroom (especially, when teaching–learning the concept of allegory), and (3) make curricular recommendations to better enable students' conceptual understandings. Such an examination will show teachers how they can move students from concrete superficial understandings to abstract mature understandings.

INTRODUCTION

"How do we know you're not making that up? Did the author really mean for you to read into the book?" Every year, several of my high-school

English students questioned how I could derive a meaning of literature that was not present explicitly. They were not convinced that *Animal Farm* could be read as an indictment of tyranny or *The Crucible* a condemnation of intolerance and hysteria. Students did not believe that an author would purposefully construct literature to say one thing but mean another. Nor did they believe it was appropriate for a reader to impose his/her interpretation on the text. They saw the reader as separate from or outside of the text, not as an agent engaging with the text. Most of my students did not fully understand the concept; nor did they understand their role in learning the concept.

According to Vygotsky, to *fully* understand a concept, one must understand the conditions of the concept's origin and development. I believe that students did not understand allegory because they did not know its conceptual development. Their lack of understanding precluded them from seeing allegory as both a literary style and a mode of interpretation, in which they should be actively involved. Ultimately, students' limited (or mis-) conception causes them to invalidate allegory and rely on a perfunctory, literal reading of the text. The purpose of this chapter is to (1) present Vygotsky's learning theory, (2) demonstrate how Vygotsky's principles can be applied in the high-school classroom, and (3) make curricular recommendations to better enable students' conceptual understandings. Such an examination will show teacher how they can move students from concrete superficial understandings to abstract mature understandings.

VYGOTSKY AND LEARNING

LEARNING: A PROCESS OF INDIVIDUAL AND SOCIAL ACTIONS

Learning is a complicated and complex process that requires action of the individual and the society in which an individual lives. Although some learning may be accomplished independently, much learning is appropriated by the learner from social interactions. This process is mediated by what Vygotsky calls more knowledgeable others (human beings with deeper knowledge than the learner) and through the use of cultural tools (which are used to transfer concepts from the external world, or society, to the internal world, or the individual). It is the role of the more knowledgeable other (e.g., teacher) to present concepts in intentional, relational, hierarchical ways to facilitate such learning. Although Vygotsky's learning theory contains several elements, this chapter will focus on the teaching and learning of *real concepts*, which Vygotsky (as cited in Gredler, 2012) defined as:

> "an image of an objective thing in all its complexity. Only when we recognize the thing in all its connections and relations [with other concepts], only when this diversity is synthesized in a word, in an integral image through a multitude of determinations do we develop a concept." (p. 122)

Vygotsky distinguished between two types of concepts—*spontaneous* or *everyday* concepts, which are highly contextual, "emerge spontaneously from the child's own reflections on immediate, everyday experiences" (as cited in Kozulin, 1990, p. 168). They can be assimilated through interaction with one's environment. Conversely, *scientific* or *theoretical* concepts "originate in the highly structured and specialized activity of classroom instruction" (as cited in Kozulin, 1990, p. 168). They are hierarchically structured and logically organized. Scientific concepts are not visibly evident and, therefore, require introduction by more knowledgeable others.

Furthermore, Vygotsky (Davydov & Keer, 1995; Kozulin, 1990) argued that scientific concepts are not automatically born of everyday concepts. That is, mere interaction with the environment (through observation or sensation) does not foster higher order thinking. Rather, higher mental processes develop through mediated activity. Similarly, Davydov (1990) asserted that theoretical generalizations, akin to scientific concepts, are obtained through mediation. He wrote,

> "… scientific knowledge is not a simple extension, intensification, and expansion of people's everyday experience. It requires the cultivation of particular means of abstracting, a particular analysis, and generalization, which permits the internal connections of things, their essence, and particular ways of idealizing the objects of cognition to be established." (p. 86)

Likewise, Schmittau (1993) explained that scientific concepts are "generally introduced in formal educational settings" (p. 30). Here, teachers can systematically mediate theoretical thinking and provide the conditions for such higher mental functions to be developed.

Vygotsky linked the development of higher mental functions with the outcomes of cognitive development and maintained that both should be the goal of education. Such higher mental functions include voluntary or self-organized attention, categorical perception, conceptual thinking, and logical memory (Gredler, 2009 & 2013). These are mental functions that allow human beings to think deeply and critically. Importantly, these higher mental functions transect all curriculum areas. Beyond theory, research indicates that educational programs grounded in theoretical learning positively affect student outcomes. With such curriculum, it was found that "student learning proceeds very quickly and with very few errors, and the knowledge mastered is meaningful and broadly transferable" (Karpov, 2013, p. 26).

CHALLENGES WITHIN THE CURRENT U.S. EDUCATION SYSTEM

Even with recorded benefits, theoretical learning has not taken root in the United States. All too often, schools emphasize the concrete or the empirical theory of cognition (Davydov, 1990; Gredler, 2009 & 2012). Concepts are presented to students as readymade. Students then take concepts as given, not knowing their origin or relation to other concepts. Another problem is that concepts are taught as fixed, unambiguous, absolute truths (Baldissera, 1993; Langer, 1989). This prevents students from seeing the evolution of the concept. Consequently, Davydov (1990) argued, children "… have not formed concepts that enable them to distinguish between superficial attributes and essential properties. They have not made the appropriate generalizations to use in organizing their world" (p. xv). Students hold incomplete concepts or entire misconceptions.

Another concern is the proliferation of content requirements (Gredler, 2009 & 2012). Over the past decade, many professional organizations and state education departments expanded the number of standards or curricular topics to be taught in each grade. This has led to what Borphy (as cited in Gredler, 2009) describes as K-12 curriculum that is a "mile-wide but an inch deep" (p. 14). This coincides with the "sprint and cover" approach, where teachers address as many concepts as quickly as possible, which presents many problems (Gredler, 2012, p. 126). First, concepts are covered poorly, superficially, without attention to their essence or universality;

second, a fast pace reduces students' opportunities to develop higher mental functions (Gredler, 2009 & 2012). As a result, students demonstrate shallow thinking. A change is needed to address these challenges in education and to better develop students' cognitive development. I do not propose a sudden and strict implementation of Vygotskian or Davydovian instructional practices, but I do believe using elements of their pedagogies could serve our schools, our teachers, and our students well.

SUGGESTIONS TO IMPROVE TEACHING AND LEARNING

Building on Vygotsky, Davydov believed the goal of teachers should be to help students transcend the formally general to the fully universal (Davydov, 1990; Davydov, Slobodchikov, & Tsukerman, 2003; Schmittau, 1993 & 1996). In other words, he supported a move from the concrete/ everyday/empirical to the abstract/scientific/theoretical. By thinking at this higher level, students could master subject-level concepts. To enhance learning, "schools must deliberately cultivate" theoretical thinking in students (Schmittau, 1996, p. 88). Teachers should help students adopt a scientific worldview, actively search for contradiction, and understand a concept's nature (Davydov, 1990; Davydov et al., 2003). Karpov (2013) describes the process of theoretical learning in several steps: first, students are taught concepts and scientific analysis procedures in specific content areas; second, students use content-related problems to internalize these concepts; third, students use additional problem-solving with newly appropriated concepts to enhance their learning. This is sometimes referred to as the "genetic method" since it is designed to "isolate mental phenomena empirically while tracing the history of their development" (Anh & Marginson, 2013, p. 149). Students are not taught just any concept. Concepts are selected carefully by teachers to position students for optimal learning.

The focus of teaching should be the essence of a concept: the essential attributes that are common in (and general to) a particular set of objects (Davydov, 1990; Davydov & Keer, 1995; Gredler, 2009). To teach the essence of a concept, one must approach cognition dialectically. In dialectical materialist theory, concepts are taught in their formation. That means, concepts are taught as processes, in a context, and in relation to other concepts and people (Anh & Marginson, 2013; Davydov, 1990; Davydov & Keer, 1995). Studying a concept in development requires an understanding of its genetic root and how its essence changes over time (Anh & Marginson, 2013; Davydov, 1990; Vygotsky in Kozulin, 1990). Dialectical materialist theory coincides with Vygotsky's cultural-historical theory, which states that learning is both sociological and psychological. Specifically, the individual cognitive processes are shaped by (mediated through) the historical and cultural systems of a particular society (Anh & Marginson, 2013; Kozulin, 1990; Wang, Bruce, & Hughes, 2011). As such, a concept is not constructed by an individual; a concept, once socially approved, is appropriated by an individual.

Gredler (2009 & 2012), building on the work of Vygotsky and Davydov, offers specific recommendations to improve curriculum and instruction. Reduce the number of terms included in curriculum materials, textbooks, and standardized exams. This can be achieved by closely examining existing curriculum resources to (1) identify and prioritize "key concepts that function as unifying threads in the subject," (2) ensure that pacing is appropriate and ample learning opportunities are available and conducive to student learning, and (3) eliminate terms "that are not addressed as concepts" (2012, p. 127). Organize true concepts into *concept networks*. These networks represent the "hierarchical and parallel relationships" among concepts in a particular subject area (2009, p. 14). Such an organizational system facilitates learning on multiple levels. Teach concepts "through other concepts that have their own internal relationships" (2009, p. 14). This demonstrates that thinking is a system of interrelated actions. Teach at a pace that allows students to develop conceptual mastery. Covering fewer real concepts more deeply and over a longer period will facilitate this. Intentionally select or design learning opportunities that are "essential to developing conceptual thinking," as these activities promote higher mental functions (2009, p. 13). Model higher mental functions for students and invite them to imitate you. According to Vygotsky (as cited in Gredler, 2009), "The learner's imitation of the model is 'a substantial factor in the development of higher forms of human behavior'" (p. 16). Request students to provide rationales when explaining and demonstrating conceptual understandings, and provide them with specific, timely feedback. Underscore the importance of speech and memory in cognitive development. Vygotsky maintained that for the child, mastery of speech was directly related to the development of thinking, and for the adolescent, "to remember is to think" (as cited in Gredler, 2009, p. 13). Many of these curricular and instructional recommendations are now commonly accepted as best practice, though translation into practice has not been swift. Such pedagogical shifts require a change in thought, change in action, and support from multiple collaborators (e.g., teachers, administrators, policy-makers, curriculum designers, professional development providers, and teacher preparation programs).

VYGOTSKY'S IDEAS APPLIED TO A HIGH-SCHOOL ENGLISH CLASSROOM

What follows is an abridged historical and conceptual analysis of "alleory." The analysis has been condensed to best suit the focus (curricular implications) of this chapter. It illustrates how Vygosky's cognitive theory, specifically the teaching and learning of scientific concepts, can be applied to a concept often taught in English class.[i] It is followed by a presentation and discussion of clinical interviews and a potential plan for instructional changes. By extension, this same teaching–learning process can be used across discipline areas.

A HISTORICAL AND CONCEPTUAL ANALYSIS OF ALLEGORY

The concept of allegory, as applied to literature, changes throughout history and is dependent upon the society in which it is used. Derived from *allos* ("other") and *agoreuein* ("to speak in the marketplace"), allegory is a form of "other speaking" where the words and meanings do not match (Allegory, 2003; Dawson, 1992; Krieger, 1981; Quilligan, 1979; Van Dyke, 1985). Ancient allegory came into being if an extended

---

[i] It is important to note that an abundance of literature on European and American allegory exists, though little has been circulated about allegory used in other parts of the world. While this does not mean that allegory has not been used by other cultures throughout history, it has shaped the Judeo-Christian examples referenced in this chapter.

metaphor was continued throughout a narrative as in Plato's *Allegory of the Cave*. Often, this included personification (e.g., gods of Greek and Roman mythology were personifications of moral principles or natural forces) (Mazzeo, 1978; Preminger, 1965). Ancient allegory also employed allusion, hence the notion of allegory containing a hidden meaning (Barney, 1979). Many allegories of this time included elements of irony, riddle, and fable and were typically considered religious or moral (Barney, 1979; Fable, 2008; Preminger, 1965; Rollinson, 1981).

The Medieval Period marked a shift in the meaning of allegory. Allegory began to expand to include multiple levels of interpretation and to embrace more than religious texts. As a didactic form, its meanings and lessons were conveyed to audiences through secular works of literature such as "The Romance of the Rose" (which teaches audiences about the Art of Love) and "The Divine Comedy" (depicting man's spiritual journey through life) (Barney, 1979; Fable, 2008; Mazzeo, 1978; Van Dyke, 1985; Whitman, 1987).

Allegory of the Renaissance also was used didactically. As an instrument of instruction, allegory was meant to disseminate "to educate gentlemen for the better management of government" (Pendergast, 2006, p. 134). Beyond imparting religious teachings, it was meant to reveal universal understandings common to man. This extended use for allegory corresponded with the philosophy of humanism and the search for self-awareness, features of the Renaissance (Murrin, 1969).

During the Modern Era (which encompasses the Baroque, Neoclassical, Enlightenment, Romantic, and contemporary periods), allegory became increasingly ambiguous and subjective. Romantics used allegory to dismantle many typical divisions between ideas (good vs. bad, right vs. wrong, man vs. nature), approaching dialecticism (Allegory, 2003; Fable, 2008; Madsen, 1996). Poststructuralists used allegory as a subjective art with open, rhapsodic, and autobiographical features, following no prescribed pattern or model (Fable, 2008). During the 20th century, spurred by Freud's publication of *Interpretation of Dreams*, the psychological allegory emerged (Fable, 2008; Madsen, 1996; Preminger, 1965). As opposed to the past, when classical allegory was used to preserve culture and history, modern allegory critiques social and political institutions, acting as the voice of dissidence (Madsen, 1996). Though some features of allegory remain constant throughout history, many forms and uses for allegory change, reflecting nuances in times and cultures.

ALLEGORY AND VYGOTSKY'S SOCIOCULTURAL THEORY

During the 20th century, Vygotsky developed the cultural–historical theory. A central principle is the sociocultural influence on psychological processes. That is, different social experiences result in different knowledge and different cognitive processes. Davydov (1990) built upon Vygotsky's theory arguing that to truly understand a concept, one must study the concept from its genesis, noting how it develops throughout history and in different contexts. A study of allegory proves just the following: the meaning and use of a concept are dependent upon the time and society in which it is used.

Many literary scholars agree that allegory changed in response to society's needs. Boys-Stones (2003) wrote that the reader's discovery of allegory "is more likely to be determined by culturally induced expectations than by any personal perspective" (p. 153). In this way, external interpretations are adopted and made internal. Dawson (1992) suggested that "allegory is not so much about the meaning or lack of meaning in texts as it is a way of using texts and their meanings to situate oneself and one's community with respect to society and culture" (p. 236). As such, allegory is used as a tool to position individuals in society. Both scholars emphasized the timely and communal nature of allegory: its relevance to a particular context, a context shaped by and serving the needs of its culture and society.

Allegory can be a difficult concept to grasp precisely because it is contoured to a given sociohistorical period. If students treat the concept in positivistic terms, believing that allegory has one definite meaning and use, they will misconstrue its nuances and development. Through mediation by the teacher, students must come to understand that allegory "never produces a 'definitive,' much less 'perfected,' text. It rather achieves various states of equilibrium, adjusting to uneven and overlapping pressures and constantly susceptible to disequilibrium and readjustment" (Whitman, 1987, p. 10). In short, "allegory serves different functions in different periods and communities" (Clifford, 1974, p. 6). This is the conceptual challenge for students to grasp.

CLINICAL INTERVIEWS: STUDENTS' UNDERSTANDING OF ALLEGORY

METHODOLOGY

To further examine my question about student thinking, I designed a small exploratory study. Specifically, I conducted clinical interviews, modeled after Piaget (Novak & Gowin, 1984). My research question asked: What are students' understandings of allegory? Participants (undergraduate students at a North Eastern public university) were purposefully selected (Glesne, 2006; Maxwell, 2005). In total, seven students (two male and five female) completed the study. They ranged in age from 19 to 24 years and differed in their majors, including human development, nursing, business, and English. (Throughout this paper, pseudonyms are used to respect the privacy of participants.) Semi-structured interviews occurred in-person and by telephone, lasting approximately 30 minutes. I posed five core questions to each participant and asked follow-up and probing questions relevant to the individual's response. Interviews were recorded, transcribed, read, and coded (multiple times). Codes emerged from the data and gave shape to the thematic organization of findings. I also used memos in an analytic way to make sense of data (Maxwell, 2005).

FINDINGS

How do you find meaning in literature?

Participants' responses fell into one of two categories: rely on self and rely on author. Five of the seven participants explained that they found meaning in literature by connecting with the text. Meaning-making was an activity that required engagement from the reader. When participants could relate to the text, connect literature to their prior experiences, they were better able to identify themes and understand the story. Jake found meaning by "applying literature to [his] own circumstances" (personal communication, May 9, 2008). Steve made sense of a text by "drawing parallels between real life and the text" (personal communication, May 9, 2008). Sally found meaning by "relating literature to [her] personal life" (personal communication, May 12, 2008). In each case, the participant interacted with literature to understand it.

Pleasure and intrigue also influenced a participant's sense-making. Two participants reported that enjoying literature, finding it entertaining and thought provoking, prompted them to invest in the text. Sally claimed, "I can easily find meaning when I read for entertainment … my attention is captured and I'm pulled into the story." Alison was not only pulled into an intriguing story but also motivated to consult outside sources. When faced with a novel idea or new style of writing, she was inspired to conduct additional research, which she used to "figure out the face value and deeper meanings of literature" (personal communication, May 9, 2008). Whether by relating the text to prior experiences or by simply enjoying the story, many participants relied on themselves to find meaning in literature.

However, other participants explained that they relied on the author's actions to comprehend literature. Sam commented that syntax (personal communication, May 9, 2008), Jill that symbolism (personal communication, May 12, 2008), and Kathryn that characterization (personal communication, May 15, 2008) influenced their understanding of a text. Alison stated, "Authors write in certain styles to convey certain emotions or ideas." Meaning, therefore, is conveyed by authorial intent, not merely a reader's interpretation. Several students believed that authors scaffold meaning by using particular grammatical structures and literary devices.

Two additional themes pertaining to meaning emerged. First, meaning-making is not a one-sided act. Two students argued that interpretation is solely dependent neither on the reader nor on the author. Instead, meaning occurs when the efforts of both the reader and the author are combined. As Alison stated, "Understanding is a creative relationship between the reader, author, and text." Second, meaning is not necessarily easy to find. Two students discussed how teachers helped them develop the skills necessary to find meaning. Sally and Jill expressed difficulty and frustration with interpreting literature. It was through explicit teaching and modeling by their English teachers that they gained confidence in reading literature on multiple levels. Jill explained it was especially helpful when teachers chunked literature into smaller sections and lead students "slowly through a dissection of the text." This suggests that finding meaning is an acquired ability.

What is the reader's role/responsibility in finding/making meaning in literature?

All participants agreed that the reader had to be active to grasp the meaning of literature. Students defined activity in different ways. For some, like Steve and Kathryn, it meant paying attention to the structure of the text. Syntax, diction, characterization, and organization revealed textual understandings. For others, like Alison and Jill, it meant gaining knowledge and understanding new ideas. For Sally and Lauren, it meant connecting literature to personal experiences. Regardless of how meaning was found, all participants had similar comments as Kathryn: the reader must "choose to be actively engaged … to pay attention and to analyze."

How many meanings does a single piece of literature have?

Participants viewed the number of meanings a given text has through two different lenses: as determined by the reader or by the author. All students believed that literature has myriad meanings. Phrases like "meaning is not set in stone" (Alison), "a text can be deciphered in many ways" (Jill), "there is an array of different meanings" (Sally), and "there is an infinitesimal amount of meanings" (Kathryn) occurred frequently throughout interviews. Although in agreement about the multiplicity of meanings within a text, participants did not concur on the derivation of such interpretations.

Five students believed that the reader greatly influences a text's number of meanings. This includes prior life experiences, ability to see things from different perspectives, current cognitive level, use of imagination, familiarity with an author's style, and ability to draw parallels with other literature (Alison, Jill, & Lauren). These participants saw meaning as subjective, determined by the audience. Two students believed that the number of meanings is dependent upon the purpose and style of the author. Consequently, a text has a predetermined number of meanings infused by the writer (Steve & Jake). To these students, literature had commonly accepted interpretations. They offered that misunderstanding occur when the reader takes more out of or puts more into the text than the author intended (Alison & Jake).

How would you describe allegory?

Four participants (Steve, Sally, Lauren, & Jill) did not know how to describe allegory. Of the four who did not explain allegory, three said they were unfamiliar with the concept. Steve, however, acknowledged hearing the concept before, referring to allegory as "an English term." Two participants (Alison and Jack) understood the classical concept of allegory: "other speaking." They described allegory as "a story that has additional deeper meaning" (Jake) and "something that conveys something at face value but also has a deeper broader meaning" (Alison). One participant (Kathryn) had a classical and modern understanding of allegory. Kathryn acknowledged the literal and allegorical levels of literature; moreover, she pointed out that allegory "uses a concrete situation (through characters, events, and objects) to illustrate an abstract, intellectual idea."

What is an example of allegory?

The four students who did not explain allegory did not provide examples of allegory. Three students, however, did elaborate on at least one illustration of the concept. Alison cited the Bible as an example of allegory. Her rationale was that the Bible could be interpreted "literally and scientifically ... that morals and broad meanings were conveyed to the audience ... [and] that it combined the past and present societies." Jake mentioned *The Chronicles of Narnia*. He argued that it was allegorical because it "illustrated man's moral quest (good versus evil) ...[and] paralleled the Bible (the king fights for his believers)." Kathryn commented on *The Allegory of the Cave*. She explained that it conveyed universal understandings: the "objects passing behind the man represented the ideas of humanity."

DISCUSSION

A highly developed understanding of allegory may be dependent on context. There seems to be an association between the participant's major and his/her concept of allegory. The three students who had never heard of allegory were nursing or business majors. The four students who had various understandings of allegory (limited, classical, and modern) were human development or English majors. Kathryn, the English major, had the most complex and comprehensive understanding of the concept. This suggests that allegory (as a genre or interpretive mode) may be considered content specific and designated to the English curriculum. Moreover, since most (four out of seven) undergraduates could not explain allegory, the concept may also be time specific. Perhaps it is taught to particular audiences at a specific point in education, for example, college English majors. This could explain why non-English major undergraduates had an incomplete or nonexistent concept of allegory.

Students who had a more developed concept of allegory also had a more developed understanding of how meaning is found in literature. Rather than seeing interpretation as reader-bound or author-bound, they conceived meaning-making as dialectic. Kathryn, Jake, and Alison all explained that interpretation exists in the juncture of personal experience and authorial intent. Moreover, they argued that literature does not have an unlimited number of meanings precisely because a text is constrained to the author's purpose and reader's engagement. These same students also elaborated on reader engagement more than other participants did. Simply knowing literary techniques was not enough to find meaning in literature. Kathryn, Jake, and Alison discussed the importance of focused attention, analysis, and additional research. This requires purposeful action from the reader.

While some participants innately seemed to know how to engage with a text, others learned from their teachers. Sally and Lauren mentioned several techniques their teachers used to help them refine interpretation skills. Their teachers explicitly taught concepts and modeled their application, chunked literature into manageable bits, drew parallels among texts in an author's line of work, and made connections between various works of literature. That is, teachers taught students techniques for finding meaning in literature.

Meaning does not automatically spring forth from literature. A person does not mindlessly find allegory in a text. The act of interpretation, of finding meaning, requires engagement with and operations on a text. Engagement can be fostered by providing students literature that is interesting, entertaining, and thought provoking. Engagement can also be fostered by helping students see relevancy in the text. When students feel more connected to the text, they are more likely to work on understanding the text. Additionally, students can be taught to interact with literature: to acknowledge personal subjectivity, to use literary techniques that aid analysis, to consider authorial intent. Furthermore, students should be taught that literature, specifically allegorical interpretation, is dialogic. As Alison concisely summarizes, allegory is "grounded but not set in stone."

CURRICULAR AND INSTRUCTIONAL IMPLICATIONS: A BETTER UNDERSTANDING OF ALLEGORY

THEORETICAL GROUNDING FOR IMPROVED CURRICULUM

As previously stated, Davydov (1990; Davydov & Keer, 1995) argued that the state of education prohibits students from truly comprehending a concept. He contends, "an essential obstacle to improving curricula and teaching methods is the narrowly sensationalistic and naturalistic conceptions of learning that have not yet been overcome in pedagogy" (Davydov, 1990, p. 325). Such conceptions are primarily developed through ready-made empirical or everyday experiences and are limited to concrete reductionist meanings. Consequently, a student "learns" a concept as fixed or given, regardless of its development or complexity. To improve learning, teachers must help students develop theoretical thinking and appropriate scientific concepts. According to Vygotsky (Kozulin, 1990) and Davydov (1990; Davydov & Keer, 1995), scaffolding a deep understanding of a concept requires several conditions: the teacher must mediate learning; the concept should be taught from its genesis and through its development, with emphasis on its dialectical nature; the concept should be connected to prior knowledge; instruction should begin with the abstract; and the student should be actively engaged in the appropriation of new or refined understandings.

The conditions Vygotsky puts forward for optimal teaching and learning are often at odds with today's classrooms. Understanding a concept in its dialectical and historical–cultural sense takes time. This is antithetical to "current K-12 curricula [which] are described as 'mile-wide but inch deep, trivial pursuit, or parade of facts'" (Brophy, as cited in Gredler, 2009, p. 14). All too often, the push is to move onto the next unit, cover another set of terms, rather than develop deep, abstract understandings of scientific concepts.

A PLAN FOR IMPROVEMENT

My proposed curricular changes pertain to 11th- and 12th-grade English. The new curriculum goes beyond the information and content included in typical literature anthologies, to include supplemental materials from each major historical period. It involves collaboration with social studies, music, and art teachers. It promotes cooperative learning and reflective thinking. It incorporates multiple and varied assessments. The curriculum entitled "Allegory in Flux: From Its Genesis to the Present" is designed for a 10-week period.

BEGINNING WITH THE ABSTRACT

Hegel (Davydov, 1990) argued that instruction should begin with the abstract. He asserted, "true abstraction and a real universal (rather than a sensory concrete or the formally general) must be put as the basis of instruction oriented toward forming comprehending thought" (Davydov, 1990, p. 319). Moreover, he believed that abstract thoughts should be presented to students early in their academic career, that students not be held at the sensory level. Consequently, I propose to introduce the universal element of allegory, "other-speaking," early in a student's academic career. This scientific concept could be introduced when students have reached the "fluent" developmental stage of reading. In the fluent stage, students can effectively read a variety of texts, become involved in literature, and comprehend at a deeper level. They employ strategies for finding meaning, summarizing details, and thinking critically. Fluent readers have transitioned from learning to read to reading to learn (Developmental, n.d.; Developmental, 2007).

THE IMPORTANCE OF PRIOR KNOWLEDGE

Not all information is spontaneously learned through sensory experiences. In fact, scientific concepts are not directly observable and, therefore, cannot be appropriated through everyday experiences (Davydov, 1990; Davydov et al., 2003; Vygotsky, in Kozulin, 1990). Instead, theoretical thought must be cultivated by connecting new information with prior knowledge. Davydov (1990) referred to this as correlating the past and present (which will ultimately be connected to the future). The notion of prior knowledge is not new. Over the past several decades, it has been researched in numerous contexts and found to have a positive relationship with learning. For instance, prior knowledge has been shown to increase the rate for constructing new knowledge, decrease prediction errors during the learning process, and enable learners to appropriate complex concepts (Williams & Lombrozo, 2013). To activate students' prior knowledge, I recommend starting instruction with Aesop's fables. Since many students are familiar with Aesop's fables, they would have a current psychological scheme on which the teacher could build elaborated and nuanced understandings of allegory. Then, I suggest progressing to ancient Greek, Medieval, Renaissance, and finally modern allegorical literature.

FROM ITS ORIGIN   THROUGH ITS DEVELOPMENT

Such a chronological approach to allegory also permits students to see how the concept changes over time. Rather than presenting the concept as objective, I would teach allegory in its ambiguity and in its cultural–historical context, as it continually comes into being. To do so, I would use more than literature; I would incorporate history, culture, religion, and the arts to situate each work that we studied. I would collaborate with teachers outside of the English department to plan the unit. For example, I would consult the music and art teachers for period-specific allegorical music, paintings, pictures, and movies. I would also try to schedule the unit so that the study of allegory in American literature correlated with themes and concepts in the U.S. History class. As A.Z. Red'ko (Kozulin, 1990) found in his study of historical concepts of 5th- to 7th-grade students, "historical reality is highly complex, contradictory, and dynamic. Analyzing and explaining its particular events presupposed consideration of many factors in their internal interconnection, in their development" (p. 154). By presenting allegory from an historical and interdisciplinary approach, I could show students (1) how the concept has developed over time and (2) how it has both shaped and been shaped by culture.

MEDIATION BY THE TEACHER

Hegel, Vygotsky, and Davydov (Kozulin, 1990) considered mediation central to human reason. This is because mature psychological processes operate with and on scientific concepts, concepts that are not sensorially apparent to a person and, therefore, require introduction by an external entity. Vygotsky (Kozulin, 1990) argued that human beings are one source for such mediation. Furthermore, Vygotsky and Davydov purported that knowledge is not independently constructed, but rather socially approved and then assimilated by the individual. As Gal'perin posited, "all of man's mind is specified for him from without; he appropriates it" (Davydov, 1990, p. 314). In the realm of education, teachers can serve as mediators. To mediate the appropriation of scientific concepts, teachers must first understand psychology, cognition, and content-material. Then, they must select and present "material whose mastery, from the outset, assures their [students'] development of content-based abstractions, generalizations, and concepts" (Davydov, 1990, p. 341). Following their advice, I have purposefully selected materials to help students understand allegory's essence and evolution.

LITERARY CONTENT OF THE CURRICULUM

Content for the curriculum is divided into three categories: literary, rhetorical, and sociohistorical. "Literary" denotes the allegorical works studied. The unit on allegory would begin with several Aesop's fables, such as *The Wolf in Sheep's Clothing, The Ant and the Grasshopper,* and *The Hare and the Tortoise.* Here, students would note the use of personification and the didactic nature of each fable. As an ancient form of allegorical writing, fables are limited in their levels of interpretation, typically suggesting one message in addition to the literal meaning. I would move on to ancient myths like *Echo and Narcissus, Zeus' Lovers*, and excerpts of the Odyssey. At this point, students would notice how the myths, as allegories, are developing into longer stories and how they are moving away from characterization of animals to gods, goddesses, and mortals; yet the myths are still culturally bound and used to teach morals. Biblical parables, for example, *The Mustard Seed, The Good Shepherd, and The Prodigal Son*, also would be studied. Students then would understand that parables embody the quality of extended metaphor and maintain allegory's didactic tradition by conveying universal truths. Parables are meant to be read on multiple levels, with particular emphasis on moral, character, and spiritual development.

Once students grasped allegory in its ancient forms, I would proceed to examples of Medieval and Renaissance allegories. I would use excerpts from Part I of Guillaume de Lorris' *The Romance of the Rose* and selections (Book IV: Friendship and Book VI: Courtesy) from Edmund Spenser's *The Faerie Queene*. Here, students would detect the presence of ancient allegory and the emergence of new allegorical forms during this epoch. For example, both works continue to reference classical myths and scripture; however, they grow in complexity. *The Romance of the Rose* and *The Faerie Queene* represent allegory as a narrative (i.e., literature with a developed plot that includes action, elaborate settings and characterization, and a climax). Moreover, they signify a shift in levels of interpretation and secularization of allegory. Allegory no longer is limited to two meanings (the literal and the religious). Secular ideas (e.g., history, philosophy, politics, and sexuality) become focal features of allegorical works. Furthermore, allegory of this time illustrates more creative liberty as authors are seen as artisans with individual imagination, not merely instruments for divine communication.

Next, students would examine allegorical essays and short stories of the early Modern period. Works such as Swift's *A Modest Proposal*, Hawthorne's *The Minister's Black Veil*, and Poe's *The Masque of the Red Death* would be studied. Students would identify how allegory is morphing into a creative and subjective or individualized genre and mode of interpretation, no longer bound to religion. Although these allegories continue to say one thing and mean another, they are now more complex. They are also darker than allegories of the past, presenting topics of economic depression, famine, murder, death, and despair. Early modern allegories continue to convey messages, but in a critical or satirical manner, as opposed to a spiritual manner.

The final 4 weeks of the unit would be spent analyzing Miller's *Death of a Salesman*. Like its immediate predecessors, contemporary allegory maintains the feature of "other-speaking" while adopting a more ambiguous form. Interpretation is determined not only by the author but also by the reader. Set in a current time, contemporary allegory's topics and themes are readily accessible (even personally experienced) by its readers. Consequently, the audience brings with them unique experiences that shape their interpretations of the literature. Students would recognize this ambiguity in (for instance, the paradox of human existence, Willy's death, the American Dream) and the sociohistorical influences on (for example, the psychological tones of Freud and the mechanization of work and living) *Death of a Salesman*.

RHETORICAL AND SOCIOHISTORICAL CONTENT OF THE CURRICULUM

"Rhetorical" indicates the dialogue and debate about allegory by philosophers and rhetoricians. Students will not grasp all elements of allegory simply by studying literature. For example, students will not see the dispute over allegory as a genre or mode of interpretation by reading *The Romance of the Rose* or *Death of a Salesman*. In addition, students need to analyze essays, letters, and critiques of literary scholars. I would use primary documents from Heraclitus, Plato, Augustine, Dante, Coleridge, Frye, Fletcher, and Quilligan to provide students with a deeper understanding of the concept.

"Sociohistorical" signifies the cultural influences on and/or products of allegory. If an interdisciplinary approach to teaching the concept of allegory were not possible, I would take it upon myself to include historical and cultural information during English class. During each period, we would study political, cultural, economic, and religious movements. I would also incorporate the visual and performing arts. We would examine paintings such as *Allegory of Music* by Filippino Lippi, *Allegory of Queen Elizabeth*, and *Allegory of Age* by Titan. We would study music like Sibelius' *New World Symphony (Symphony No. 2)*, Sondheim's *Into the Woods*, and McLean's American Pie. We would probe movies such as *The Wizard of Oz, A Space Odyssey,* and *The Lion, the Witch, and the Wardrobe*. Students would analyze each film on its literal and allegorical level looking for symbolism, personification, and allusion. This learning would be connected to students' current understanding of allegory in literature.

STUDENT ENGAGEMENT

Kant (Davydov, 1990) noted, "'thinking' means 'acting'" (p. 250). Bibler (Davydov, 1990) hypothesized that theoretical thought was marked by mental experimentation. Davydov (1990) argued that a scientific concept functioned as "a mental activity by means of which an idealized object and the system of its connections ... are reproduced" (p. 249). Il'enkov (Dayvdov, 1990) suggested that children should "repeat the discoveries of human beings" (p. 320). The common theme among the aforementioned theorists is activity. Current theorists and researchers have elaborated on this notion of learner agency. They have studied behavioral, emotional, and cognitive components of what is termed "school engagement" and found that engagement is associated with positive learner outcomes (Li & Lerner, 2013). Students must be actively engaged in thoughtfully planned and mediated learning experiences. During the unit on allegory, students will complete several meaningful activities.

For meaningful learning to occur, students must "relate new knowledge to relevant concepts and propositions they already know" (Ausubel in Novak & Gowin, 1984, p. 7). One way to scaffold meaningful learning is using concept maps, which provide a framework for hierarchically organizing concept meanings (Novak & Gowin, 1984). They have the ability to "counteract the superficial treatment of concepts occasioned by the failure to develop a coherent curriculum that identifies essential concepts and probes them in sufficient depth" (Schmittau, 2004). Concept maps will be used throughout the entire unit. At the end of each period (classical, Medieval, Renaissance, early and late Modern), students will be required to complete one concept map. All concept maps must have at least two main branches: meanings and literary devices or techniques. Concept maps will be created individually. Then, students will have the opportunity to share their maps in small groups. Collaboration will provide students the opportunity to see a peer's conceptual understanding of allegory and revise or expand upon his/her own understanding. Concept maps are also useful as assessment tools. As the teacher, I can utilize them to gauge students' grasp of key ideas. Concepts maps will also be used longitudinally.

One of two culminating activities in the allegory unit is the *Here's What!—So What?—Now What?* heuristic (Lipton & Wellman, 2004). This activity requires large chart paper divided into three columns. It would be used to highlight allegory's development throughout history. In small groups, students would use their concept maps to complete the *Here's What!* column. They would list specific facts or information about meaning and literary devices for each of the allegories studied. Then, in the *So What?* column, students would interpret the data. This requires students to record patterns, note inconsistencies, make inferences, and draw conclusions. Finally, students would complete the *Now What?* column. Here, students discuss implications of their learning, make predictions, or pose questions for further study. In this case, the *Now What?* column would serve as a springboard to the second culminating activity.

Ultimately, students will compose, share, and reflect on their own allegorical piece, be it prose, drama, verse, or lyrics. Students will write an allegorical composition, including a précis that explains how their work qualifies as an allegory. Then, they will share their piece with at least three other students. The audience will lead a discussion about the author's piece, explaining their interpretations and commenting on how meaning was shaped. Afterward, the author will write a reflection on the entire allegory unit. This journal will illustrate his/ her understanding of interpreting, writing, and situating allegory in a context. The selection of appropriate content and facilitation of meaningful learning activities will better enable students to fully understand the concept of allegory.

The curricular and instructional recommendations just described extend beyond the English classroom. These same principles can be applied to classroom across discipline areas and grade levels.

CONCLUSION

Though strong in many ways, today's U.S. education system faces several challenges. One of these challenges, and the focus of this chapter, is developing habits of theoretical learning. In today's fast-paced world and in an era when quantity is valued over quality, too often teaching and learning are discussed in terms of efficiency and mass market. Many people have the misconception that teaching and learning mean covering as many concepts as possible as quickly as possible, without attention to the depth, nuance, or development of concepts. In this process, more focus is often—and unfortunately—paid to the teaching of information rather than the teaching of human beings. Vygotsky calls for "(a) replacing the tacit assumption about the educational process from a goal of focusing on content to restructuring the learner's mental functions and (b) analyzing concept in the curriculum in a new way" (Gredler, 2012, p. 128).

Vygotsky believed that a central goal of teachers should be to help students develop higher mental functions, which students could then generalize, enabling them to regulate their own learning. But such a process is not automatic; hence, the necessary mediation of teachers. It is our responsibility as teachers to help students understand that learning is not neat, static, or hastily acquired. By teaching allegory as a scientific concept, I can help students see allegory from its genesis through its development. Through this "new 'model' ... dialectical, theoretical thought" takes root (Davydov, 1990, p. 172). Students become critical thinkers, actively engaged in the learning process—not simply parrots of static information.

Indeed, curriculum can be created (through mediation of the teacher and by student engagement in meaningful learning activities) that enables students to better understand abstract concepts. My intent is to move students from an empirical, superficial understanding of concepts to a richer, nuanced understanding. In this space, students should examine scientific concepts forming, in particular, social-historical contexts. They should identify a concept's core essence. They can construct new, more complex understandings by seeing concepts in relation to other concepts, and through a dialectial relationship, continue to reconstruct more sophisticated understandings of concepts. Teachers of all subject areas and grade levels can make adjustments in curriculum and instruction to help students develop higher mental functions.

KEYWORDS

- allegory
- conceptual understanding
- curriculum
- dialectical relationships
- learning theory
- Vygotsky

REFERENCES

1. Allegory in Literary History. (2003). *The dictionary of the history of ideas*. Retrieved from http://etext.lib.virginia.edu/cgi-local/DHI/ot2www-dhi?specfile=%2Ftexts%2Fenglish%2Fdhi%2Fdhi.o2w&query=allegory&docs=div1&title=&sample=1-100&grouping=work.
2. Anh, D.T.K., & Marginson, S. (2013). Global learning through the lens of Vygotskian sociocultural theory. *Critical Studies in Education*, 54*(2)*, 143–159.
3. Baldissera, J.A. (1993). *Misconceptions of revolution in history textbooks and their effects on meaningful learning*. Proceedings of the Third International Seminar on Misconceptions and Educational Strategies in Science and Mathematics, Ithaca, NY, Misconceptions Trust.
4. Barney, S.A. (1979). *Allegories of history, allegories of love*. Hamden, CT: Archon Books.
5. Boys-Stones, G.R. (2003). The stoics' two types of allegory. In G.R. Boys-Stones (Ed.), *Metaphor, allegory, and the classical tradition* (pp. 189–216). New York: Oxford University Press.
6. Clifford, G. (1974). The Transformations of Allegory. Boston, MA: Routledge & Kegan Paul.
7. Davydov, V.V. (1990). Types of generalization in instruction: Logical and psychological problems in the structuring of school curricula. In J. Kilpatrick (Ed.), *Soviet Studies in Mathematics Education*, Volume 2. Reston, VA: NCTM.
8. Davydov, V.V., & Keer, S.T. (1995). The influence of L. S. Vygotsky on education theory, research, and practice. *Educational Researcher*, 24*(3)*, 12–21.
9. Davydov, V.V., Slobodchikov, V.I., & Tsukerman, G.A. (2003). The elementary school student as an agent of learning activity. *Journal of Russian and East European Psychology*, 41*(5)*, 63–76.
10. Dawson, D. (1992). *Allegorical readers and cultural revision in ancient Alexandria*. Los Angeles, CA: University of California Press.

11. Developmental Stages of Reading. (n.d.). Sarasota County School. Retrieved from http://www.sarasota.k12.fl.us/sarasota/develstgrd.htm#Fluent%20Stage%20 (DRP%20Levels%2041+).
12. Developmental Stages of Reading. (2007). Ansbach Elementary School. Retrieved from http://www.ansb-es.eu.dodea.edu/Developmental%20Stages%20of%20Reading.htm.
13. Fable, Parable, and Allegory. (2008). *Encyclopedia Britannica.* Retrieved from http://www.search.eb.com/eb/article-9110449.
14. Glesne, C. (2006). *Becoming qualitative researchers: An introduction,* 3rd edition. New York: Pearson.
15. Gredler, M.E. (2009). Hiding in plain sight: The stages of mastery/self-regulation in Vygotsky's cultural-historical theory. *Educational Psychologist, 44(1),* 1–19.
16. Gredler, M.E. (2012). Understanding Vygotsky for the classroom: Is it too late? *Educational Psychology Review,* 24, 113–131.
17. Karpov, Y.V. (2013). A way to implement the neo-Vygotskian theoretical approach in the schools. *International Journal of Pedagogical Innovations, 1(1),* 25–35.
18. Kozulin, A. (1990). *Vygotsky's psychology: A biography of ideas.* Cambridge, MA: Harvard University Press.
19. Krieger, M. (1981). The two allegories. In M.W. Bloomfield (Ed.), *Allegory, Myth, and Symbol* (pp. 355–370). Cambridge, MA: Harvard University Press.
20. Langer, E.J. (1989). *Mindfulness.* Cambridge, MA: Da Capo Press.
21. Li, Y., & Lerner, R. M. (2013). Interrelations of behavioral, emotional, and cognitive school engagement in high school students. *Journal of Youth & Adolescence, 42(1),* 20–32.
22. Lipton, L., & Wellman, B. (2004). *Pathways to Understanding: Patterns and Practices in the Learning-Focused Classroom,* 3rd edition. Sherman, CT: MiraVia.
23. Madsen, D.L. (1996). *Allegory in America: From Puritanism to Postmodernism.* New York: St. Martin's Press.
24. Maxwell, J.A. (2005). *Qualitative Research Design: An Interactive Approach,* 2nd edition.
25. Thousand Oaks, CA: SAGE.
26. Mazzeo, J.A. (1978). Allegorical interpretation and history. *Comparative Literature,* 30(1), 1–21.
27. Murrin, M. (1969). *The veil of allegory: Some notes toward a theory of allegorical rhetoric in the English renaissance.* Chicago, IL: University of Chicago Press.
28. Novak, J.D., & Gowin, D.B. (1984). *Learning how to learn.* New York: Cambridge University Press.
29. Pendergast, J. (2006). *Religion, allegory, and literacy in early modern England, 1560–1640: The control of the world.* Great Britain: Ashgate.
30. Preminger, A. (Ed.). (1965). Allegory. *Princeton encyclopedia of poetry and poetics.* Princeton, NJ: Princeton University Press.
31. Quilligan, M. (1979). *The language of allegory: Defining the genre.* Ithaca, NY: Cornell University Press.
32. Rollinson, P. (1981). *Classical theories of allegory and Christian culture.* Pittsburgh, PA: Duquesne University Press.

33. Schmittau, J. (1993). Vygotskian scientific concepts: Implications for mathematics education. *Focus on Learning Problems in Mathematics*, 15(2–3), 29–39.
34. Schmittau, J. (1996). Cognitive development. In J.J. Charnblis (Ed.), *Philosophy of Education: An Encyclopedia* (pp. 85–89). New York, NY: Garland.
35. Schmittau, J. (2004). *Uses of concept mapping in teacher education in mathematics.* Paper presented at First International Conference on Concept Mapping, Pamplona, Spain.
36. Van Dyke, C. (1985). *The fiction of truth: Structures of meaning in narrative and dramatic allegory.* Ithaca, NY: Cornell University Press.
37. Wang, L., Bruce, C., & Hughes, H. (2011). Sociocultural theories and their application in information literacy research and education. *Australian Academic & Research Libraries,*42*(4),* 296–308.
38. Whitman, J. (1987). *Allegory: The dynamics of an ancient and medieval technique.* Cambridge, MA: Harvard University Press.
39. Williams, J.J., & Lombrozo, T. (2013). Explanation and prior knowledge interact to guide learning. *Cognitive Psychology,* 66(1), 55–84.

CHAPTER 5

CROSSING THE BARRIERS, EXPANDING KNOWLEDGE, FOSTERING RELATIONSHIPS: TEACHER PREPARATION PARTNERING WITH COMMUNITY ORGANIZATIONS

JOHN DELPORT,[1] JAMIA RICHMOND,[1] NANCI E. HOWARD,[1] and TALYA KEMPER[2]

[1]*Coastal Carolina University, 100 Chanticleer Dr E, Conway, SC 29528, United States*

[2]*California State University, 401 Golden Shore, Long Beach, CA 90802, United States*

ABSTRACT

This chapter will review a special education teacher preparation program's implementation of a field experience, that spanned formal (schools) and informal educational settings (recreation programs and non-school-based programs), over a semester-long course with a seminar component. From this chapter, readers will gain insight into the literature and rationale for working with community organizations and programs beyond the schools. A qualitative analysis of the data will be presented, highlighting several important themes regarding the importance of field experiences in informal learning environments.

INTRODUCTION

All teachers must revel in the world outside the classroom community ...
in education, we set about solving educational problems as if they exist in
a vacuum (Delpit, 2006, p. 93)

The narrow focus and associated restricted array of solutions to address student needs has been a consistent critique in the field of special education (Harry, 2008; Kalyanpur & Harry, 1999; Trent, Kia, & Oh, 2008; Sailor & Skritic, 1996; Skritic, 1995). This limited attention is a result of the field's emergence from the medical and behavioral psychology perspectives with their singular focus on the identification of symptoms and their associated treatment (Sailor & Skritic, 1996; Skritic, 1995). Scholars argue that in order to address the needs of all students with disabilities and their families, the focus and related solutions need to broaden in scope to a more ecological perspective, which incorporates a socio-political-cultural perspective inclusive of families and communities (Ford, Obiakor, & Patton, 1995; Harry, 2008; Trent, Kia, & Oh, 2008; Webb-Johnson, Artiles, Trent, Jackson, & Velox, 1998).

The purpose of this chapter is to highlight field experiences in community-based settings as a best practice for preservice special education teachers (hereafter referred to as preservice teachers). Therefore, we will explore and discuss how one special education preservice teacher program attempted to broaden preservice teachers' vision of practice with children with autism spectrum disorder (ASD) through providing an additional field experience in a community organization that serves such children. This experience is in addition to their prescribed program of internships and student teaching in formal school settings.

RATIONALE FOR COMMUNITY-BASED FIELD EXPERIENCES

We educators set out to teach, but how can we reach the worlds of others
when we don't even know they exist (Delpit, 2006, p. xxvi)

Students in grade 1–12, including those identified as having special needs, spend the majority of their waking hours (based on 16 hours a day) in informal learning environments, within their community (Stevens, Bransford, & Stevens, 2005). These children spend the majority of their time in non-school settings, time that is critical in their development. Informal learning environments could include, but are not limited to, afterschool programs, recreation centers, churches or other religious settings, vacation destinations, and homes. This is especially critical in relation to children with special needs and their families as these settings very often provide much needed services and resources. These non-school settings include community organizations that provide recreational programs and/or other services for students, as well as supports to families. Many families rely on these external service providers and organizations, in addition to schools, to help with the delivery of services to their children. The rapid increase in the number of students being identified as having ASD—1 out of 68 (Baio, 2014) —means that all teachers, including special educators in all types of classroom assignments, will work with students on the spectrum and their families. Serving as a valuable resource, teachers need to be aware of these organizations and programs in the community so that they can connect families with them and to work with staff members from these organizations, when possible, to develop effective programming.

The placement of preservice special educators in community-based organizations (CBO) by a special education teacher preparation program was driven by the call in the special and general education literature to increase preservice special education teachers' experiences in the community. The literature supporting this call is presented in this chapter as well as how the partnership emerged and developed between a CBO and a special education teacher preparation program. The CBO in this partnership was SOS Healthcare Inc (SOS). SOS is a CBO that serves children on the autism spectrum and their families. The partnership began with the creation of volunteer opportunities at SOS and later developed into a required community-based placement that included university-based course work. The partnership began in 2012 and the structure and related course expectations changed over time based on student, SOS staff, and faculty members' input. The preservice teachers engaged in field experiences in a variety of programs ranging from recreation programs to applied behavior analysis (ABA) therapy. The development of the partnership, as well as the outcomes of the experiences for preservice teachers, faculty, and the program, is discussed at length in this chapter.

FOUNDATIONS AND DEVELOPMENT OF THE PARTNERSHIP

The Council for Accreditation of Educator Preparation (CAEP, 2015), which governs accreditation of teacher education programs, highlights the importance of preservice teachers having field experiences in non-school settings. Standard One in their overview of standards for teacher preparation programs states,

> "High quality clinical experiences are early, ongoing, and take place in a variety of school and community-based settings .... These experiences integrate applications of theory from pedagogical courses or modules in P-12 or community settings and are aligned with the school-based curriculum." (p. 7)

The idea of community-based placements as a site for teacher learning is not a new concept. In 1948, a report commissioned by the American Association of Colleges of Teacher Education (AACTE), *School and community laboratory experiences in teacher education*, also known as "The Flowers Report," asserted the need for fieldwork experiences both in and outside of school, particularly those that provide "guided contact with children and youth of differing abilities and of differing socio-economic backgrounds" (Flowers, Paterson, Stratemeyer, & Lindsey, 1948, p. 26). These important teacher education bodies and the similar sentiment they call for have been consistent over 75 years and support the importance and need for community-based placements for teacher learning.

As the opening quote by Lisa Delpit (2006) appeals to all teachers to revel in the world outside of the school walls, the report by Flowers and colleagues (1948) recommends that programs offer opportunities for teachers to "study the community to better understand learners' needs and backgrounds ... work cooperatively with other educational agencies in the interest of children" (p. 27). In the 1950s and 1960s, in an effort to prepare teachers to work in desegregated schools and as part of the civil rights movement, scholars argued for community-based fieldwork aimed at improving teachers' understanding of the lives of children and groups they had not previously taught (Fruchter, 2007).

Community–school relationships have consistently been shown to improve school outcomes for children from historically marginalized communities, specifically those from non-dominant racial, cultural, economic, and language communities (Abrams & Gibbs, 2000; Adger, 2001; Fruchter, 2007; Honig, Kahne, & McLaughlin, 2001; Jones, 1992; Lopez, Kreider, & Coffman, 2005; Warren, 2005; Warren, Hong, Leung Rubin, & Uy, 2009). In recent years, teacher education programs have once again looked toward community-based placements to augment work in the university and K-12 school settings (Gallego, Rueda, & Moll, 2005; McDonald, Tyson, Brayko, Bowman, Delport, & Shimomura, 2013). While the models vary, sometimes vastly, in a number of ways, one commonality they share is their objective to expose teacher candidates to diverse groups of students (Sleeter, 2008). For some programs, such exposure is the central purpose of the fieldwork, while others aim to promote one or more of the following areas: assets-based thinking, a community orientation to teaching, learning about diverse children and communities, a broader conception of learning, and an ethic of service. One of the key purposes of community-based placements is its use as a tool for the dispositional learning or change of teaching candidates' thinking and orienting themselves to a more inclusive approach to teaching. Specifically, a body of research has found that the placement of preservice teachers in CBOs broadens their perspectives of the practice of teaching and provides a deeper, more contextualized, understanding of the children and families they will eventually serve (Bondy & Davis, 2000; Boyle-Baise, 1998; Burant & Kirby, 2002; Gallego, 2001; Gallego, Rueda, & Moll, 2005; McDonald & Zeichner, 2008; McDonald et al., 2013; Seidl & Friend, 2002; Zeichner & Melnick, 1996).

In special education teacher preparation literature, the bulk of community-based placements for preservice teachers falls under the banner of service learning. Service learning integrates community-based student projects into the more formal curriculum of their teacher preparation program (Serow, Calleson, Parker, & Morgan, 1996). It is not volunteering, per se, but a method of blending service and learning with the goal of academic learning on the part of the preservice teacher (Dunlap, Drew, Carter, & Brandes, 1997). Critical to the teacher education literature that addresses service-learning pedagogy is that the students provide a service. In most cases, this is an agreed-upon project between the constituents of the organization and the preservice teachers.

Originally defined in the National and Community Trust Act of 1993, the definition of service learning expanded with the reauthorization of the Serve America Act of 2009. The framework of service learning emphasizes experiences described by Pritchard & Whitehead in Muwana & Gaffney, (2011, p. 22)

(i) under which students learn and develop through active participation in thoughtfully organized service experiences that meet actual community needs and that are coordinated in collaboration with school and community;

(ii) that is integrated into and enhances the students' academic curriculum or the educational component of the community service program in which the participants are enrolled;

(iii) that provides students with opportunities to use newly acquired skills and knowledge in real life situations in their own communities; and

(iv) that provides structured time for students to reflect on their service experience.

The community-based placement discussed in this chapter aligns with the service-learning model laid out in points (i)–(iv) above, but has some key differences. One key distinction in the community-based placement in the partnership discussed in this chapter is that there was not a focus on traditional service defined in feature (i) above. Considering the terms of the first feature of meeting community needs, our student participants' activities were not geared toward meeting identified needs of the organizations constituencies in the form of a mutually agreed-on project. Rather it served as an additional field experience for the preservice teachers to provide more "hands on deck" at the placements and to provide supplementary supervision and structure for children with ASD. The focus of this community-based field experience was for the preservice teachers to learn more about children with ASD, to better understand the resources in the community, and increase their skills and knowledge as future special education teachers.

As for the second requirement of enhancing student academic progress, ii above, although our participating preservice teachers spent significant time observing and learning from the staff at their placements, they were not expected to lead or create activities, although several of the pre-service teachers did lead or create activities in their placements once they were comfortable. Some of these projects included leading activities in the recreation and after-school programs with art and games, leading a social skills session, or engaging in an ABA session under lead therapists' direction. They were essentially observers and the goal was to provide the pre-service teachers with the opportunity to "… come out from behind the shield of 'teacher' and all that it implies: power techniques, manager, advocate and placed them back in the role of the learner … without the distraction of having to assume responsibility for those they observed" (Szepkouski & Dunn, 1997, p. 7).

The community-based placement allowed them an opportunity to focus on getting to know the children and the organization free from the prescribed roles of "teacher" as occurs in a school-based placement. It also allowed them to be free of an agenda of a service project, typical of service-learning placements, and allowed them to have a deeper observation experience with more time to engage with the children. The community-based placement in this partnership allowed for the teachers to engage in a "deeper level of analysis and critique" (Szepkouski & Dunn, 1997, p. 7) than school-based and service-learning placements typically allow. They did learn from this experience, but the learning was personal and not directed by the special education faculty or the SOS staff. The preservice teachers learned strategies at SOS, like providing a direction and engaging in planned ignoring, which they were able to implement during their time in the public school field experience.

A unique feature of this partnership and its development over 2 years was that the preservice teachers were able to reflect on their learning from university-based classes, their time in the public schools that they were placed in as part of the program of study as well as their community-based placement. Although the university-based class was a foundational class, participants were able to compare the content they learned in the university-based class about children with ASD and what they actually observed in the community-based placements. Additionally, they were able to transfer learning from their time in SOS to the school setting. Some even had the opportunities to work with children at SOS and at their school-based placement.

In regards to the fourth feature, from the Serve America Act of 2009, a core objective was for participants to reflect through a guided reflection process after each session. In these reflections, they connected their experiences, observations, as well as knowledge and skills learned in relation to their students and how these might inform their school-based placements they were in during this experience as well as past and future school-based experiences. Additionally, and perhaps more importantly, was how they connected these experiences to their role as a special education teacher in the future.

The first feature of the 2009 Act "... meet the actual community needs ..." was not directly met with this experience, but the focus was on how these experiences would allow them to be engaged in that kind of work in the future.

The primary goal of the faculty for the preservice teachers was that of learning—learning about the children with identified needs outside of school and the university classroom, about organizations that serve these students and families, and lastly, knowledge and skills from experts in these organizations that work with children with ASD and their families in the community. This goal was strengthened by the placement of preservice teachers in a well-established and structured organization, their observations, and constant guided reflection with culminating individual reflection projects. The ultimate goal of this community-based placement was that future teachers would engage with this type of organization as well as connect students and families with them and ultimately have a more nuanced understating of children, families, and communities beyond the university classes and school-based settings.

Anderson, Swick, and Yff (2001) point out that the studies on service learning in teacher education highlight the influence of the experience on dispositions and skills such as "commitment to teach, a caring attitude, and acceptance of diversity" (Cited in Mayhew & Welch, 2001, p. 209). They put forward that more research need to be conducted on how service learning enriches content knowledge and its ability to help teachers develop knowledge, skills, and attitudes to better serve families and students. Although not strictly a service-learning experience, this community-based placement aimed to provide an opportunity, specifically in the classroom discussions, weekly reflections, and final project, to reflect on the experiences they had to develop the following dispositions--commitment to teach, a caring attitude, and acceptance of diversity. The later three constructs are associated with service learning in the literature (Muwana & Gaffney, 2011).

COMMUNITY-BASED PLACEMENTS AND SERVICE LEARNING IN SPECIAL EDUCATION TEACHER PREPARATION

There is an evident dearth in the review of the special education teacher preparation literature in relation to the placement of preservice special education teachers in non-school or community placements. Within this small body of literature, the majority of the literature specifically addressed service learning. Only one study (Adams, Bondy, & Kuhel, 2005) more closely aligned to the community-based placements described in this chapter. Mayhew and Welch (2001), in their call to special education teacher educators to engage in service-learning experiences with preservice teachers, point out that little research exists on the use of service learning in special education teacher preparation programs. This section will provide a brief overview of the literature that has engaged in service-learning experiences and other community-based placements in the preparation of preservice special education teachers.

Welch and James (2007) investigated the impact of reflections on students engaged in a service-learning experience in local classrooms that served students with special needs. The participations were from two sections of an introduction to special education class (n = 26) placed in a local classroom for students with special needs. The focus of this study was on the preservice teachers reflections. The researchers posited that "reflection is a critical element in the pedagogy of service learning" (p. 276).

Smith (2003) as a requirement of a service learning class with a field experience component had university students, from variety of majors, partnered with a peer from the local school district enrolled in a transition program for students over 18 years old. The university students in this study were partnered with a student in the transition program between the ages of 18 and 21. The transition program focused on providing work experience, independent living, and functional skill development for the program students. The university students were required to meet with transition program students and work with them on the transition-related activities like working in community business, self-care activities, and other activities to increase the programs participants' functional skills.

The university students were expected to write reflections about their experiences as well as identify, and conduct, transition-related activities with their peer partner from the transition program. Analysis of the university students' reflections showed increases in their knowledge of the lives of individuals with disabilities, as well as an expanded understanding of the needs and desires of individuals with disabilities.

Jenkins and Sheehy (2009) reviewed the use of service learning in their special education undergraduate program over a 9-year period. These service-learning placements ranged from local adult disability service agencies to schools and programs serving school-aged children with disabilities. They found that the service-learning opportunities that they utilized in their special education teacher preparation program had a positive impact on preservice special education teachers' learning. This was evidenced in the findings that the experiences helped these preservice teachers develop a broader vision of practice and service in relation to teaching and toward individuals with disabilities. Evidence to support this was found in the analysis of the preservice teachers' reflections as well as their pre- and postscores on a measure of attitudes and perceptions toward people with disabilities.

Novak, Murray, Scheuermann, and Curran (2009) developed a service-learning experience with in a special education course on collaboration and consultation. Students in the class partnered with parents and local community organizations serving children with special needs. Parents of children with disabilities were included in the weekly session of this semester-long class. They provided direct instruction as well as guidance in discussions and activities within the classroom at the university. The authors posit that having the parents in the class gave the university students first-person narratives of the parents' experiences and also direct feedback on the utility and accuracy of the service-learning projects.

Muwana and Gaffney (2011) in their class *Culture of disabilities across the lifespan* incorporated a service-learning component. The service-learning component and class work focused on advancing community services for individuals with disabilities. Although the class was offered out of the special education department, it is not clear if the students were special education majors as they were all freshmen and may not have declared their major. In addition to the university students, six community partners from four agencies and three consumers from these agencies were included in the research and development of the service-learning projects. University students completed pre- and postsurveys as well as were interviewed by the researchers. The community partners and three of the four consumers were interviewed and all completed a satisfaction questionnaire. Several focus groups were also conducted with a mix of university students, community partners, and consumers. The community partners and the consumers positively rated the service-learning projects with disabilities in the focus groups, interviews, and satisfaction questionnaires. The community stakeholders reported that they felt included and had a valuable role in advancing the service-learning projects. The students in the interviews and focus groups shared that they had learned skills and knowledge that were important to their future endeavors and interactions with people with disabilities. Positive changes were also found in the pre- and postsurvey of the students' attitudes toward individuals with disabilities. Overall, the data analysis showed that all stakeholders—students, faculty, community organization leaders, and consumers with disabilities—felt that the experience was important and positive.

Adams, Bondy, and Kuhel (2005) deviated from the prescribed feature of service learning. In their study, participants did not complete a "service project" agreed upon by the organization and students, but rather spent time in an existing reading program in an urban housing project, which is more in line with the experiences of the preservice teachers of this study. Like the majority of the studies already discussed, the participants were not predominantly special education majors.

All of these studies, except for Adams, Bondy, and Kuhel (2005), focused on the use of service learning and included at least several of the following characteristics: a focus on shared service projects agreed on by organization and students, guided reflections, and deliberate and explicit embedding of service learning in the university-based classed session. Most of the research were not specific to preservice special education teachers and relied on school-based placements versus community-based placements. Furthermore, those that did include preservice special education teachers' did not have their knowledge and skill development for their future classrooms as a central component but rather their dispositional changes. That is, the focus was based on how the students reflected on their perceptions toward working with individuals with disabilities and had little inclusion on knowledge and skills for their future classrooms. This does not always translate into concrete knowledge and skill development that allow students to better work with and serve children with disabilities and their families. The research of Jenkins and Sheehy (2009) is important in understanding the importance of using service learning and non-traditional field placements for preservice special education teachers in that they looked at longitudinal data within the special education department with special education majors.

DEVELOPMENT OF THE COMMUNITY-BASED PARTNERSHIP PROJECT

THE SPECIAL EDUCATION PROGRAM

Coastal Carolina University (CCU) is a midsize liberal arts university in the South East region of the United States. The college offers four undergraduate degree programs and several postgraduate programs. The college of education offers several professional programs, and preservice teachers in the undergraduate programs are required to complete the university core, which typically takes 2 years, and then apply to the professional program. Students are required to pass the Praxis I test, have a minimum cumulative grade point average (GPA) of 2.6, and also have passed all the education prerequisites with a C grade or higher. Students complete the 2-year professional program and leave with teaching endorsements and a Bachelor of Arts degree in the area they were enrolled in, in this case Special Education or Early Childhood. The preservice teachers participating in this partnership over the first three semesters (from Fall 2012 through Fall 2014) were all enrolled in the special education teacher preparation program (special education majors) in the college of education. During the spring 2015 semester, the preservice teacher participants were enrolled in either the early childhood teacher preparation program (early childhood majors) or the special education teacher preparation program (special education majors).

THE PARTNERS

SOS is a grassroots community organization, conceived in 1989, that serves the city of Myrtle Beach and greater Horry County in South Carolina. The primary focus of SOS is on providing services, resources, and recreation programming to students with ASD and their families. They offer a wide array of services for students and families touched by ASD as well as other developmental disabilities from birth to adulthood. These services include an Autism Clinic (ABA therapy and case management), recreation/social programs, adult transition services, and summer camps (SOS Healthcare, Inc.; http://scautismhelp.com/). Through the development of the partnership, preservice teachers have been placed in a variety of these programs. We will provide a brief description of these.

- Friday Knights II is a recreation program for children with ASD and serves two groups of children organized by age—kindergarten through 8th grade and 9th grade through to young adults. This program, a sister program to the original Friday Knights out of upstate New York, is described as a "Recreation and Family Support Program for Children with Autism Spectrum Disorders or Difficulties in Relating" (http://scautismhelp.com/programs/friday-knights/). Sessions are typically 2 hours in length and are held once every other week over the school year.
- SOULS is an after-school social skills group, which was developed as a "Social and Life Skill Building Program for Autism Spectrum Disorders or Difficulties in Relating" (http://scautismhelp.com/programs/s-o-u-l/). Like Friday Knights II, there are two groups, one for elementary-aged children and another for older children and young adults. Sessions for SOULS are typically 2 hours in length and are held once a week. SOSs after-school program follows the local school districts calendar and is offered 5 days a week in the afternoon once schools are out. This is described on the website as follows: "Designed for children with Autism, this program provides homework help and fosters social skills. The program will also include other planned activities such as wellness/fitness groups, games, arts and crafts, and gardening" (http://scautismhelp.com/programs/after-school-program/).
- SOS provides ABA services, which are offered under the Building Futures Autism Clinic program, at three of their clinic locations as well as in home depending on the families' needs. The preservice teachers participate in the individual ABA therapy sessions, held at the main clinic. The Building Futures Autism Clinic is described on their website as providing, "… a complete skill assessment with the ABLLS-R, verbal behavior training, social skills training, play/

leisure skill development, functional assessment of problem behavior, as well as parent and staff workshops" (http://scautismhelp.com/programs/building-futures/).

- Making Change Consignment Store is a transition and job-training program for young adults with autism. Its purpose is to, "… offer adolescents and young adults with Autism Spectrum Disorders the opportunity to gain vocational skills through training for work in a retail environment. The participants will create products to sell in the store as well as interacting with customers and improving their social skills."

METHODOLOGY

PILOTING THE STUDY

Over the past 2 years of implementing these community-based placements in partnership with SOS, expectations have changed, as well as the courses coupled with it. The number of sessions as well as required assignments have been edited and adapted based on the information received each semester. The expectations and changes to the experiences are described here.

During the Fall 2012 semester, the CCU special education faculty arranged a meeting with SOS to explore the potential of having preservice teachers volunteer with SOS programs. SOS agreed to let the preservice teachers volunteer in their programs. SOS had already worked with interns from other departments at CCU. A plan was developed to have preservice teachers enrolled in a Foundations of Special Education class, EDLD 370, to participate in the Friday Knights II program. EDLD 370 was a requirement for all preservice teachers in the special education program. During the first 3 weeks of class, students were invited by the instructor of EDLD 370 to volunteer for the Friday Knights II program. Attendance was low during this stage of the partnership and we believe this can be attributed to it being voluntary as well as there being no integration into the course.

Prior to the Spring 2013 semester, changes were made in relation to the placement of preservice teachers in SOS programs. Students from Fall 2012 (EDLD 370) provided informal feedback and confirmed our belief that attendance was low due to it being voluntary and having no class expectations attached to it. The faculty and SOS staff also met, prior to the start of the spring semester, and discussed the issue of low attendance and buy-in. During the Spring 2013 semester, the changes were implemented to increase student participation. Special Education majors were enrolled in EDLD 469, a 2-days/week field experience with a focus on special education in the middle grades and a weekly seminar. This field experience was a program requirement. The students in this course were required to participate and attend Friday Knights II or SOULS every other week. To increase preservice teachers' buy-in to the community-based placement at SOS programs, the seminar was only scheduled to meet on campus every other week versus once a week. The alternate week was scheduled for attending the SOS program, and the total number of hours was what would be expected without the community-based placement. Attendance was required with a total of seven sessions to be completed by the end of the semester. In addition to requiring attendance, a guided reflection was to be completed after each session as well as a final reflection project and the grade from these were included in the overall class grade. During this semester, students' attendance increased significantly compared to Fall 2012.

FEEDBACK FROM THE PILOT STUDY

Prior to the Fall 2014 semester, university faculty and SOS staff met to discuss the partnership and community-based placement opportunities. This meeting confirmed that SOS staff felt that the experience was valuable in that they could have an increased influence on teacher training and how these future teachers, many of who work in the local school district, will work with the children they serve. They also reported that having the students on site provided some additional help with the management of the programs activities and having "extra set of hands and pairs of eyes." As a result of this discussion, the after-school program and the ABA therapy sessions were included as options for the preservice teachers. Attendance, number of sessions, as well as the reflections, and projects were also discussed. Even though attendance was required in the Spring 2013 experience, the faculty and SOS staff discussed that the students seemed to struggle, as a whole, to attend all seven sessions due to work commitments and other obligations. It was decided that for the Fall 2014 experience, a minimum of four sessions would be required and that the reflections and final reflection project would hold more weight in the form of points toward their total class grade. SOS staff discussed that the preservice teachers needed more prompting from the special education faculty to increase their engagement and interaction with the children in their respective placements. In addition, SOS staff attended the first week of class to describe the programs, the children served in the SOS programs, and provided an overview of ASD.

IMPLEMENTING THE STUDY, FALL 2014–SPRING 2015

Incorporating the suggestions from the feedback meeting, a new class of preservice special education teachers enrolled in the Foundations of Special Education course (EDLD 370) and were placed in SOS programs. Students were expected to attend a total of four sessions over the semester, complete a guided reflection after each session, and complete a final reflection project. Students completed their community-based placements in Friday Knights II, SOULs, the SOS after-school program, or ABA therapy.

The most recent of the experiences in this partnership were completed during the Spring 2015 semester. The class was EDSP 200 Q, also a Foundations of Special Education class like EDLD 370. EDSP 200 Q was different from the previous class in this partnership in that it not only included preservice special education teachers but early childhood preservice teachers as well. Students were expected to attend a total of seven sessions over the semester, an increase from the previous semester. The SOS staff also attended the initial class sessions, as was done in the Fall of 2014 semester, to describe the programs, and focus on the kinds of activities they wanted to the students to engage in. They spent significant time describing ASD as well as providing the concrete examples of where the university students could engage with children while in their experiential learning. The option to work with young adults with ASD, at the Making Change Consignment store, was also included as a community-based placement option. The number of students in Spring 2015 was over 50 compared to Fall 2014 that had less than 20 students. Due to this increase in students, other organizations and opportunities outside of SOS were included as placement sites for community-based placements. These included other-area after-school programs serving students with disabilities, shadowing a school psychologist, and working in the university gym with a student with Down syndrome.

DATA ANALYSIS

This was a phenomenological qualitative study of the preservice special education teachers' experiences while in their community-based placements in several SOS programs and how they perceived these through their guided reflections, final projects, and interviews. Creswell (2013) describes the phenomenological method as a study that "describes the meaning of the lived experiences of several individuals about the concept of phenomenon" (p. 51). The participants were placed in all the programs discussed that SOS offers. The Fall 2014 preservice teachers completed a total of four sessions but several completed above this required amount. The Spring 2015 preservice teachers completed at least seven sessions, while some completed more than this.

DATA SOURCES

Qualitative analysis of phenomenological data sources used students' narratives in the form of reflections, final projects, and conducted interviews. A total of 105 reflections were included in the analysis, 54 from Fall 2014 and 61 from Spring 2015. The reflections were completed after each session and students followed a reflection guide (See Appendix A). These reflections varied in length from half a page to three pages with the average being one page. Participants completed between three and seven reflections of their experiences. A total of 33 final projects were included in the analysis with 17 participants completing a paper between five and ten pages and 16 completing multimedia presentations of which the majority were PowerPoints. The final projects were completed once the preservice teachers had completed their required sessions (see Appendix B for a description of the final project). Due to time limitations at the end of the Spring 2015 semester and students leaving campus for the summer, interviews were only conducted with the Fall 2014 participants. A total of seven participants volunteered to be interviewed. The interview questions were adapted from interview questions used in the research conducted by Bondy and Davis (2000), Boyle-Baise (1998), Burant and Kirby (2002), and Gallego (2001). The questions, a mix of open ended and directed questions, are included in the appendices of this chapter. The first three authors conducted the interviews over the phone as well as in person with the preservice teachers. When the interviews were complete, they were transcribed by graduate assistants and were then analyzed as laid out in the section below.

As noted, data collected included transcribed interviews of the Fall 2014 preservice teachers, as well as weekly reflections and final projects from Fall 2014 and Spring 2015 participants. Table 5.1 provides a descriptive overview of the participants, program placement, as well as the data sources and number of resources in each category. Table 5.2 provides an overview of the community-based placements, university-based classes, and assignments related to the community-based placement.

The data and analysis included in this chapter are drawn from the Fall 2014 and Spring 2015 community-based placements. The participants included in the spring 2015 were only those preservice special education teachers and excludes the preservice early childhood teachers to allow for consistency across data sources. A total of 33 preservice special education students participated in providing data for this study. There were 15 preservice teachers from Fall 2014 and 18 from Spring 2015. The majority of the students were female, and Caucasian, which is typical for a preservice teacher education program (Sleeter, 2008).

TABLE 5.1   Study Participants, Program Placements, and Data Sources

	Fall 2014	Spring 2015
Total Participants	15	18
Males	2	1
Females	13	17
Programs		
Friday Knights II	5	5
Afterschool	6	5
SOULs	4	5
ABA Therapy	—	—
Other	—	3
Data Sources		
Reflections	54	61
Final Project		
• Multi Media	5	11
• Paper	10	7
Interviews	7	0

TABLE 5.2 Overview of Community-Based Placements and Related University Course and Assignments

Semester	Community-Based Placements	Preservice Teachers Major Area	University Class	Assignment/Expectations
Fall 2012	Friday Knights II	Special Education	EDLD 370 Foundations of Special Education	Volunteer
Spring 2013	Friday Knights II SOULs	Special Education	EDLD 469 Middle Level Special Education Field Experience Seminar	Attend at least seven sessions, write a reflection after each session, complete a culminating project
Fall 2014	Friday Knights SOULs SOS, Afterschool program ABA therapy	Special Education	*EDLD 370 Q Foundations of Special Education Students were also in a Elementary-Level Special Education Field Experience	Attend at least four sessions, write a reflection after each session, complete a culminating reflection project
Spring 2015	Friday Knights II SOULs ABA therapy SOS, Afterschool Program Making change Other afterschool programs in the area	Special Education and Early Childhood	**EDSP 200 Q Foundations of Special Education	Attend at least seven sessions, write a reflection after each session, complete a culminating project

*The Q designation is part of a university-wide improvement plan to increase the experiential learning opportunities for students, and this class was granted the status after applying for it in Spring 2013.

**The Special Education program changed from a Learning Disabilities focus to multicategorical. All course prefixes changed from EDLD to EDSP but the content in EDLD 370 and EDSP 200 remained the same.

DATA CODING

The researchers conducted three rounds of analysis of the data sources during which data were organized into themes and categories. A constant comparative technique was used to identify themes and increase the rigor. The first three authors read through each of the data sources independently of each other and conducted a thematic analysis using the reflection questions to find themes inductively. Each data source was then analyzed for any other phenomena that did not come out during the initial analysis. Once these themes had been identified, the themes from each data source were compared and were included or excluded based on thematic match (Creswell, 2007). After this final independent round of analysis, the first three authors met and discussed the themes they found and analyzed for agreement and disagreement on these themes. Several themes were consistent across researchers and were included in the findings; several were not included due to there not being appropriate agreement. Six themes emerged from the data, each will be discussed next.

FINDINGS

In the analysis of the data sources, six themes emerged and were supported within participants' responses and across different data sources (e.g., interviews, session reflections, and final projects). The themes reflect the preservice teachers' perceptions in relation to their development of conceptual and practical knowledge while being in the community-based placements working with students with ASD. The six themes that emerged were as follows:

1. Gaining valuable experiences from experts in the field
2. Seeing children beyond the school walls in a nonacademic environment
3. The individuality of children labeled as having ASD and/or other disabilities
4. Connecting theory to practice
5. Development of dispositions
6. The overall experience

In this section, we will present the themes that emerged from the analysis of the data. Italics have been added to several of the excerpts for emphasis. All names are pseudonyms to protect the preservice teachers' identities.

GAINING VALUABLE LEARNING EXPERIENCE FROM EXPERTS IN THE FIELD

Lisa, a participating student, described the value of working with experts from the field of ASD in a community in the excerpt below. In it she describes how she was not familiar with how to handle a specific situation with a child with ASD and relates how this observation and learning will influence her as a future teacher.

> Although I was only there for a short period of time I feel like I have learned plenty of useful information that will benefit me in the future as an educator … While there I observed one of the students become very frustrated about doing homework and go into a complete fit. *I had no idea how to handle this situation so it was great to see how the SOS team worked together to resolve the situation in a timely manner while attending to the other children.* (Lisa, Reflection 1, Fall 2014)

Jennifer shares about her experience working with an expert and how she took time to explain the strategy and her choice for using it with a particular child. Jennifer found this experience very beneficial.

> One of the therapists, Tammy, was very kind and helpful. She took her time to explain to me what she was doing with the kids and why she was doing it. She answered all of my questions and helped me understand the methods she used on the kid she was shadowing. I really appreciated the time Tammy took explaining to me because she was the first person to take her time and tell me what she was doing and why she was doing it (Jennifer, Reflection 7, Spring 2015).

Several preservice teachers describe strategies they learned while observing therapists working in the community-based placement with the students. Esther described a "great social strategy" that she will use in her classroom (Esther, Reflection 1, Spring 2015). Another preservice teacher, Lara, states:

> I have learned a lot from the staff and children there. I also was able to *talk more personally with one of the staff members.* We discussed one of the children who was having an off day. He explained that it can be hard working with a nonverbal child, especially when there is something

wrong, because you don't always know what to do to help him/her. But, he told me it is okay not to know sometimes, trial on error and showing an effort is sometimes all it takes (Lara, Reflection 3, Spring 2015).

And finally, Sandra sums up the strategies she's learned with the following statement:

I would love to take some of these ideas and add my one twist to them to help these kids really branch out ... Something I would love to try in my school someday is the charades game we played with the iPad. I always just saw that game as a fun home game but for the kids at soul [sic] it can be really useful to help them think creatively, communicate with others and take part in teamwork. They saw this game more as a reward rather than a strategy to socialize which I think is great way to get kids excited. (Sandra, Reflection 3, Spring 2015).

SEEING CHILDREN BEYOND THE SCHOOL WALLS IN A NONACADEMIC ENVIRONMENT

The second theme was related to the preservice teachers' reflecting on the value of being in this setting. This was an important finding in that it was a guiding objective of our placement of the preservice teachers in informal non-school settings. The preservice teachers' reflections centered on the way in which they could focus on being with children with no academic demands. This is a key difference between their placements in academic or school settings where they had many university academic task demands on them. They reflected on the value of this context that allowed them to focus on developing relationships with the children. Many scholars have described and written about the importance of developing relationships with children to be an effective teacher and this experience seems to have allowed the preservice teachers to develop the skills as well as practice the skill of developing relationships (Gallego, Rueda, & Moll, 2005; Kalyanpur, & Harry, 1999; Sleeter, 2008; Warren, 2005).

These are conversations *I would not have been able to have in the group setting, and I never had that personal ... like conversation with a student with autism before this experience.* I liked being in a setting where I could be a friend and *not an authority figure or teacher most out of everything I experienced.* (Cari, Final Project, Fall 2014)

Evelyn describes the nonacademic environment as fun and a better way to connect with a student. She states:

It was really fun because it wasn't structured, I just got to talk with them and find out some of their interests. I talked to one of the quieter girls who told me she liked reading. Later, when we were playing games, she wandered off into another room where there were books. Knowing that she likes reading, I started asking her questions about the types of books she likes and she started opening up to me about why she likes reading and being by herself. This was a great learning moment for me. Previously, she would just ignore me when I said hi or asked her a question, but *once I finally figured out what she liked it helped me make a better connection with her.* I also like reading, and I have heard of some of the series of books that she has read, so we had something to talk about. (Evelyn, Reflection 3, Spring 2015)

Jamie provides critical insight into how being with children in the classroom and in the community can help teachers better understand their students when they become aware of the differences in behavior and engagement of these children in school and non-school settings. Jamie worked with one child in both her school-based field experience and at the more informal SOS setting.

One interesting thing I noticed was that the child I know from field placement *acts completely different outside of the classroom,* she seems happier and none of the behavior problems that she has in school were a problem at Friday Knights. (Jamie, Reflection 1, Fall 2014)

THE INDIVIDUALITY OF CHILDREN LABELED AS HAVING ASD AND/OR OTHER DISABILITIES

The next theme is critical in two areas. The first is that scholars consistently point out that ASD is a spectrum of needs and deficits. Teachers and other professionals need to understand this and not create a monolithic approach or perception of these students. Preservice teachers consistently reflected on how they were exposed to and learned about the individuality of children with ASD. This is a critical disposition and belief for preservice teachers to develop in order to effectively work with students in their future classrooms. This also provides another way to look at students, which differs from the medical and psychological models that focus on a collection of symptoms but often erase the individual. In the first excerpt, Janice identifies the diversity within the autism spectrum and how it manifests in different individuals.

> I learned so much about children with autism through this experience. *I learned how diverse people with autism are. Not one case of autism is exactly like another.* I also *learned about different tics and stems through this experience.* (Janice, Final Project, Fall 2014)

Brandy reiterates this understanding of diversity within the autism spectrum the following semester in this statement:

> I learned that each individual has a different way they need to be cared for. *Just because they have the exact same disability does not mean that they need to be treated and cared for the same way.* (Brandy, Final Project, Spring 2015)

In her final project, Cari reveals that even though her class and field experiences gave her a basic understanding of ASD, this experience provided her with invaluable insights into how important it was to understand that all of the students on the spectrum are different.

> I learned *how different students with autism are from each other.* You learn in textbooks that children with autism have this or that characteristic or quality, and *you cannot help but make generalizations* … The personalities these students have were *more complex than I could have ever imagined off of any generalizations I have heard in a textbook and that is what I learned about students with autism.* (Cari, Final Project, Fall 2014)

CONNECTING THEORY TO PRACTICE

Teacher education programs are charged with providing concrete knowledge and skills to preservice teachers to assist them in developing a repertoire of practices to effectively serve children. One of the skills covered in a behavior management course is ABA. Below, Jessica reflects on ABA, what she learned about it in the university classroom, and how it was implemented in the nonacademic setting.

"The Building Futures Clinic taught me more about what Applied Behavior Analysis is. We went over it briefly in class but I was able to see it be put into practice … Being able to observe therapy sessions at the Building Futures Autism Clinic was extremely eye opening … ABA is completely different than being in a classroom. These therapist works on basic living skills such as potty training, picking up food with a fork, and matching. They also teach the children how to do math, tell time, how to communicate effectively and many other skills" (Jessica, Final Project, Spring 2015).

In several special education courses, services and strategies for children with various disabilities are described. Here, Esther and Jennifer connect some of the ideas discussed in class with their experiences in the community setting.

There was a student there last night that was not at the previous session, Richard, whom I chose to sit with during snack to talk to because his behavior seemed quite different than the other kids. He had some obvious behavior issues but showed no signs of "classic" autism. After everything I said to him he would reply "no one cares" which caught me off guard a little. I could tell some of the kids avoided him and I wanted to learn more about him. After speaking with a therapist working with us that night she told me he did not have Autism but rather EBD but was labeled Autistic so he could receive these services. *It was interesting to see this first hand since we have talked about it quite a bit in class.* After spending time with this child I understand because he really needs these services. (Esther, Reflection 2, Spring 2015)

Jennifer wrote

In class, *we learned about certain ways to control the children when they are having a "fit."* At Friday Knights, when a child didn't get their way or something in their little world went wrong, some would start screaming and crying and it seemed nearly impossible to get them calm. When this happened, at first we would try to calm them down, if this didn't work, we ignored them because we didn't want to give them the attention they were seeking by having this "fit." *If we had not discussed what to do in this situation, I might not know what to do when this happened at Friday Knights.* (Jennifer, Final Project, Spring)

In another course on communication and language, students learn about communication boards and picture schedules. Lara describes the process of actually using a picture schedule with a child in the nonacademic setting. She states:

> An interaction that I had with a student was helping him complete his own picture schedule throughout the afterschool program … *I chose this interaction because I thought it would be an interesting way to see how using a picture schedule actually works for a child.* (Lara, Reflection 2, Spring 2015)

As preservice teachers reflected on the practices that they observed, tried out, and have added to their repertoire, recurring comments about flexibility or not trying to control the students, behavior management, and the use of specific activities emerged as data were analyzed. With regards to flexibility, it is an important practice to be adaptable. On this point, Esther states:

> I learned that things do not and will not go as you plan them to go and you must always be prepared to change course and adapt to the needs of the children. The lady in charge had planned some activities for us to all do that the materials just did not work and we had to improvise but the staff did a great job to not letting the kids know that nothing was going as planned. (Esther, Reflection 3, Spring 2015)

Additionally, Jennifer states:

> I learned that not all the kids can be handled in the same way. You have to change the way you talk and interact with each one of them. This will help me as a future teacher because I will have to adapt my lessons and teaching methods for each student. (Jennifer, Reflection 4, Spring 2015)

Lara describes it in this way:

> I learned that I have to be able to understand the differences between how each child works. I noticed that things that worked for me last week with one child did not work for me this week with a different child. Adaption is key … Trying different methods to get a child to complete a task can work wonders. (Lara, Reflection 2, Spring 2015)

When reflecting on behavior management, several students identified specific practices or strategies they would like to apply in their future classrooms.

This is when I encountered the first of many emotional breakdowns. He became so frustrated that he couldn't fit the last piece into the correct spot he began screaming, crying, and kicking. As part of ABA, the therapist stressed to me that it is important not to give in to these behaviors. We had to make sure we did not make eye contact and not to help him out. After about a minute and a half of screaming, the child picked up the puzzle piece and slowly got it into the correct place. The therapist and myself basically "threw a party" for him. He got his reinforcers (doritos) and he got extra playtime. The therapist told me that he could do it and these tantrums always come when he gets severely frustrated. I was kind of shocked to see this happen but the way it was handled opened my eyes to many different scenarios that could have happened. Most parents or teachers would give in to the child and help him do it or maybe show him how to do it. (Jessica, Reflection 2, Spring 2015)

Esther reflects on the practice of not belittling the child in front of others.

There was one kid, the child who one of the therapists told me last session was labeled autistic so he could get the services, who really had an emotional meltdown this session. I learned from the staff how to handle that and how to deal with the other children who see what is going on. They did an amazing job not to put him down to the others by not drawing attention to it and by responding to their questions with statements like oh he's just had a bad day like we all have sometimes. (Esther, Reflection 3, Spring 2015)

Pamela describes a concrete example used while in the community organization that will influence the way she delivers instruction when she is a teacher. She specifically credits the way the activities were delivered in the community setting.

I love how the activities are centered around different learning skills. There is a verbal activity, a visual activity, a fine motor skills activity, etc. It would be nice to be able to do the same thing in the classroom when we begin to teach. (Pamela, Reflection 3, Fall 2014)

Table 5.3 displays the strategies and practices that the preservice teachers found useful from the community-based experience. These are strategies they indicated they would likely use again in their future careers as educators.

DEVELOPMENT OF DISPOSITIONS

Preservice teachers also reflected on critical dispositions needed for effective teaching. These were expressed in a change of how they felt from the beginning of the experience to the end and that the result was important for them as a teacher. For example, they reflected a gradual shift from nervousness to a level of comfort. In their initial reflections, Arie and Brad describe feelings of "nervousness" or "being nervous" as they began the experience (Arie, Reflection 1, Fall 2014; Brad, Reflection 1, Fall 2014). In the Spring 2015 data, Kourtni states: "I honestly was not very comfortable my first night because everything was very new to me" (Kournti, Reflection 1) and Evelyn felt " … really nervous going into it because I had no idea what to expect"

TABLE 5.3 Strategies Preservice Teachers Found Useful in Community Setting

Strategies, Practices, Ideas to Use in the Future
1. Games
a. Name introduction game(for making connections)
b. Blind puzzle (for communication)
c. Knot game (for teamwork and communication)
d. Charades with iPad (for teamwork and communication)
e. Getting to know you games/activities (for making connections)
2. Increased amount of one-to-one attention
3. Extra time for processing information
4. Prompts, visual cues
5. Classroom jobs and roles
6. Teaching manners and behaviors in authentic settings
7. Teaching more about Individual Education Plans (IEP) and transition plans
8. Form strong relationships with students
9. Rewards

(Evelyn, Reflection 1). Over time, the preservice teachers began to experience comfort in the community setting. In her third reflection, Janice states: "This was my favorite time so far at SOS. I am starting to get more and more comfortable with the children" (Janice, Reflection 3, Fall 2014). Kourtni sums it up this way, "Each Friday Knights I attend I get a lot more comfortable with the kids" (Kourtni, Reflection 3, Spring 2015).

They also realized they needed to have patience and understanding. Mary describes an experience where patience and understanding were critical to completing a project.

> Today we did a project with glue, glitter and spiders to make a spider web. While working with one little girl, she was very particular about which spiders she wanted to use and was insistent on only using certain ones. This is what was comfortable to her so this is what she wanted. *Having patience and understanding is key to working with children who have learning differences and special needs.* It may not make sense to anyone but this child about why she only wanted to use certain spiders, but that doesn't matter. With something as small as to using certain spiders for an art project, if possible I find it most beneficial to let the child use what she wants, rather than "fighting" her on it. Preventing an "outburst" and keeping her happy was more important in this case. (Mary, Reflection 1, Fall 2014)

After an incident at a restaurant where one student refused to get off a table and another refused to eat because it didn't look like she wanted it to, Sandra reflects on the importance of patience.

> I learned to have a lot of patience this session, it is definitely important because with situations like the one with Devon and Annie you don't want to make a scene out in public but you want them to understand, I think Adam did a good job at handling both of these situations and I tried to imagine myself in his shoes to see what I would have done and if I would've had the patience. (Sandra, Reflection 6, Spring 2015)

Many participants reflected on the need to be dependable and consistent to be more effective with the children as well as being actively engaged. Several of the preservice teachers also talked about the opportunity to practice leadership skills and the importance of these skills when teaching.

THE OVERALL EXPERIENCE

Consistently preservice teachers articulated the value of the experience overall. A feature of the experience that was described consistently was watching the experts work with the children. Evelyn describes this as one of her favorite things in the excerpt below.

> I really enjoyed so much about this experience. One of my favorite things was watching the adults interact with the children and how they stopped certain behaviors. I learned some great words to use to get the children to understand what is expected of them! (Evelyn, Final Project, Spring 2015)

The children, themselves, were a source of enjoyment for the preservice teachers. Many expressed how enjoyable it was to observe and work with these children in nonacademic settings. In describing what she enjoyed the most, Jessica states "Being able to participate in the sessions and bond with the children" and "Seeing the improvements the children made week after week" as her most enjoyable in her experience (Jessica, Final Project, Spring 2015). Linda says, "I thoroughly enjoyed seeing the kids at SOULs every week and I plan to continue to volunteer at SOULs" (Linda, Reflection 7, Spring 2015).

Many found the overall experience to be a unique and enjoyable experience where they learned a great deal about themselves, ASD, and working in a community setting. This was another critical finding to support the importance of non-school settings for preservice teachers. For example, Lisa stated, "I am glad that I had this experience and I look forward to many more experiences like this in the near future" (Lisa, Reflection 4, Fall 2014). Additionally, Pamela included her own children in this experience, and together as a family, has committed to continue learning through volunteering in this setting.

> What I enjoy most about coming on Friday nights [sic] is learning something new from these kids. Everyone thinks we are there to teach them something but honestly, they are teaching us. Whether it be a lesson in patience, understanding, or how to reinvent a typical mundane lesson, these kids are always pushing us to think differently and out of the box. I also enjoy watching the enthusiasm on my kids' faces when

they ask if we are going to SOS. They have enjoyed this experience as much as I have. Even though this was my last night for class requirements, we have made a decision as a family that we are going to continue to go and learn and have fun! (Pamela, Reflection 4, Fall 2014)

Cathy described her excitement in continuing to work in the program and the knowledge gained that she will use as a future teacher in the excerpt below. Cathy is one of the preservice teachers who chose to continue working with the Friday Knights II program after her class experience ended.

I look forward to working for the program for the rest of the semester as well as the spring semester and gaining even more knowledge to use for my teaching as well as at work. (Cathy, Reflection 4, Fall 2014)

CONCLUSION

Although the findings are promising and support the placement of preservice teachers in community-based settings, several areas will need to be addressed moving forward to create a more effective experience and in understanding the effectiveness of the experiences. In the next round of data collection and analysis, changes will occur. The first is to fine-tune the reflection questions (see Appendix A) to access additional, deeper information. In addition, when we interview (see Appendix C) the preservice teachers as well as when we provide feedback on the reflections and final projects (see Appendix B), we will ask them to help better define and explain several statements that occurred frequently. Statements such as "I gained patience," "I learned how to deal with a child with autism," and "I learned how to communicate better" need greater explanation to determine what the preservice teachers' definitions of these concepts are. In addition to this, the instructor of the course will attempt to provide preservice teachers with more opportunities to compare and contrast their experiences in the schools and community setting. The data analysis will also closely examine the comparison between school settings and CBOs in terms of behaviors of students with disabilities and these students' interactions with preservice teachers.

The literature is clear that preservice teachers benefit from well-structured and deliberate experiences in the community organizations

(Adams, Bondy, & Kuhel, 2005; Anderson, Swick, & Yff, 2001; Boyle-Baise, 1998; Burant & Kirby, 2002; Gallego, 2001). Special education programs are currently not engaging in this experience to the same levels as general education programs (Mayhew & Marshal, 2001; Novak, Murray, Scheuermann, & Curran, 2009). This chapter provided an overview of the existing literature and community-based placements as well as literature-based rationale for increasing the placement of preservice special education teachers in these settings. The program development and structure were described and will allow readers to explore this type of experience in their institution. The preservice teachers' narratives provide an important basis and insight into how the experience will ultimately help them become better teachers. Moving beyond the school walls is important and critical as we attempt to prepare teachers for the diversity and complexities of the public schools where they ultimately will be charged with being social-change agents. These kinds of experiences move them beyond the school and university setting and place them in the communities they will ultimately serve. It also teaches them to use programs outside of the classroom setting in order to help provide children and families with a network of services that include not just school but also community organizations to improve these children and families' lives.

APPENDIX A

GUIDED REFLECTION PROMPTS

You will write a short, guided reflection after each session. In this you will be expected to reflect on your experiences as well as how it could, if at all, help you as a future teacher and how it helped/hindered you in understanding children with special needs.

Below is an outline they will complete for the reflection:

(a) Briefly describe an interaction you had with a student. What happened and why did you choose this interaction?

(b) What, if anything, did you learn about the child(ren) who you worked with today?

(c) What, if anything, did you learn about yourself?

(d) Please describe something you learned that may be relevant to your working in the school or that you are going to try out in your school.

(e) Anything else you want to add (e.g., interactions with staff, families, siblings)?

APPENDIX B

FINAL PROJECT DESCRIPTION

You will complete culminating project in which you will choose from one of the options below to show the knowledge and skills you have acquired during your time in the experience.

Option 1: *Paper*

You will write a paper in which you address the questions below as well as reflect and draw on your individual session reflections.

1) What did you most enjoy about the experience? Why?
2) What did you least enjoy? Why?
3) What did you learn, if anything, about students with disabilities?
4) Did you use any skills, knowledge you learned at your program in the working with families and students and vice versa?
5) What, if anything, would you change about the experience?
6) Any other reflection? Ideas? (e.g., Are there any interactions/ learning that you feel were significant?)

Option 2: *Multimedia presentation*

You can create a poster, PowerPoint presentation, movie in which you show your learning. The questions below can help guide this as well as your session reflections.

1) What did you most enjoy about the experience? Why?
2) What did you least enjoy? Why?
3) What did you learn, if anything, about students with disabilities?
4) Did you use any skills, knowledge you learned at your program in the working with families and students and vice versa?
5) What, if anything, would you change about the experience?
6) Any other reflection? Ideas? (e.g., Are there any interactions/ learning that you feel were significant?)

APPENDIX C

INTERVIEW PROTOCOL FOR CROSSING BARRIERS INTERVIEWS

(Adapted from Bondy & Davis, 2000; Boyle-Baise, 1998; Burant & Kirby, 2002; Gallego, 2001)

1) Please describe the program you did your service learning with, for example, number of participants, gender distribution, and group mood.
2) What were the most significant things you learned from participating in this experience?
3) Has your thinking about students with disabilities changed in any way in relation to this experience? If so, why? If not, why not?
4) Please describe an interaction(s) with specific children.
5) Please classroom compare your school and community field experiences (e.g., similarities and differences).
6) To what extent has this experience impacted upon your learning in the introduction to special education course? If so, how? If not, why not?
7) To what extent has this experience impacted upon your learning and views toward being a special education teacher? If so, how? If not, why not?
8) If you could change this experience, what would you do?

KEYWORDS

- community organizations
- community-based field experiences
- field experiences
- informal educational settings
- informal environment

REFERENCES

1. Abrams, L.S., & Gibbs, J.T. (2000). Planning for school change: School-community collaboration in a full-service elementary school. *Urban Education, 35*(1), 79–103.
2. Adams, A., Bondy, E., & Kuhel, K. (2005). Preservice teacher learning in an unfamiliar setting. *Teacher Education Quarterly, 32*(2), 41–62.
3. Adger, C.T. (2001). School–community-based organization partnerships for language minority students' school success. *Journal of Education for Students Placed at Risk (JESPAR), 6*(1–2), 7–25.
4. Anderson, J.B., Swick, K.J., & Yff, J. (2001). *Service-learning in Teacher Education: Enhancing the Growth of New Teachers, Their Students, and Communities.* Washington, DC: AACTE.
5. Baio, J., (2014). Prevalence of autism spectrum disorder among children aged 8 years: Autism and developmental disabilities monitoring network, 11 sites, United States, 2010. Retrieved from http://www.cdc.gov/mmwr/preview/mmwrhtml/ss6302a1. htm?s_cid=ss6302a1_w.
6. Bondy, E., & Davis, S. (2000). The caring of strangers: Insights from a field experience in a culturally unfamiliar community. *Action in Teacher Education, 22*, 54–66.
7. Boyle-Baise, M. (1998). Community service learning for multicultural education: an exploratory study with preservice teachers. *Equity and Excellence in Education, 31*(2), 52–60.
8. Burant, T.J. & Kirby, D. (2002). Beyond classroom-based early field experiences: Understanding an "educative practicum" in an urban school and community. *Teaching and Teacher Education, 18*(5), 561–575.
9. Council for Accreditation of Educator Preparation (CAEP). (2015). *CAEP Accreditation Standards.* Washington, DC: Council for Accreditation of Educator Preparation.
10. Creswell, J.W. (2007). *Qualitative Inquiry and Research Design: Choosing Among Five Traditions* (2nd Ed.). Thousand Oaks, CA: Sage Publications.
11. Delpit, L.D. (2006). *OTHER People's children: Cultural conflict in the Classroom.* New York: The New Press.
12. Dunlap, N.C., Drew, S.F., Carter, K.G., & Brandes, B.D. (1997). *Serving to Learn: Integrating Service Learning into Special Needs Curricula.* Columbia, SC: SC Department of Education.
13. Ford, B.A., Obiakor, F.E., & Patton, J.M. (1995). *Effective Education of African American Exceptional Learners: New Perspectives.* Austin: Pro-Ed.
14. Flowers, J., Paterson, A., Stratemeyer, F., & Lindsey, M. (1948). *School and Community Laboratory Experiences In Teacher Education.* Oneonta, NY: American Association of Colleges for Teacher Education.
15. Fruchter, N. (2007). *Urban Schools, Public Will: Making Education Work for All Our Children.* New York, NY: Teachers College, Columbia University.
16. Gallego, M.A. (2001). Is experience the best teacher? The potential of coupling classroom and community-based field experiences. *Journal of Teacher Education, 52*(4), 312–325.

17. Gallego, M.A., Rueda, R., & Moll, L. C. (2005). Multilevel approaches to document-ing change: Challenges in community-based educational research. *Teachers College Record, 107*(10), 2299–2325.
18. Harry, B. (2008). Collaboration with culturally and linguistically diverse families: Ideal versus reality. *Exceptional Children, 74*(3), 372–388.
19. Honig, M.I., Kahne, J., & McLaughlin, M.W. (2001). School-community connec-tions: Strengthening opportunity to learn and opportunity to teach. In V. Richardson (Ed.), *Handbook of Research on Teaching*. Washington, D.C.: American Educational Research Association.
20. Jenkins, A., & Sheehey, P. (2009). Implementing service learning in special educa-tion coursework: What we learned. *Education, 129*(4), 668.
21. Jones, B.A. (1992). Collaboration: The case for indigenous community-based orga-nization support of dropout prevention programming and implementation. *Journal of Negro Education, 61*(4), 496–508.
22. Kalyanpur, M., & Harry, B. (1999). *Culture in Special Education: Building Recipro-cal Family-Professional Relationships*. Baltimore, MD: P.H. Brookes.
23. Lopez, M.E., Kreider, H., & Coffman, J. (2005). Intermediary organizations as capac-ity builders in family educational involvement. *Urban Education, 40*(1), 78–105.
24. National and Community Service Trust Act of 1993, Pub. L. 103-82, Stat. 785 (1993).
25. Mayhew, J., & Welch, M. (2001). A call to service: Service learning as a pedagogy in special education programs. *Teacher Education and Special Education: The Journal of the Teacher Education Division of the Council for Exceptional Children, 24*(3), 208–219.
26. McDonald, M., Tyson, K., Brayko, K., Bowman, M., Delport, J., & Shimomura, F. (2013). "Innovation and impact in teacher education: Community-based orga-nizations as field placements for preservice teachers." *Teachers College Record* 113.8 (2011): 1668–1700.
27. McDonald, M., & Zeichner, K. (2008). Social justice teacher education. In W. Ayers, T. Quinn, & D. Stovall (Eds.), *The Handbook of Social Justice in Education*. Philadelphia: Taylor and Francis.
28. Muwana, F.C., & Gaffney, J.S. (2011). Service-learning experiences of college fresh-men, community partners, and consumers with disabilities. *Teacher Education and Special Education: The Journal of the Teacher Education Division of the Council for Exceptional Children, 34*(1), 21–36.
29. Novak, J., Murray, M., Scheuermann, A., & Curran, E. (2009). Enhancing the prepa-ration of special educators through service learning: evidence from two preservice courses. *International Journal of Special Education, 24*(1), 32–44.
30. Sailor, W., & Skritic, T.M. (1996). School/community partnerships and educational reform: introduction to the topical issue. *Remedial and Special Education, 17*(5), 267.
31. Serve America Act of 2009, Pub. L. No. 111-13, Stat. 1460. (2009). Retrieved from http://www.nationalservice.gov/about/legislation/edward-m-kennedy-serve-amer-ica-act.
32. Seidl, B., & Friend, G. (2002). Leaving authority at the door: Equal-status community-based experiences and the preparation of teachers for diverse classrooms. *Teaching and Teacher Education, 18*(4), 421–433.

33. Serow, R.C., Calleson, D.C., Parker, L., & Morgan, L.E.I.G.H. (1996). Institutional support for service-learning. *Journal of Research and Development in Education, 29,* 220–225.

34. Skritic, T.M., Ed. (1995). *Disability and Democracy: Reconstructing (Special) Education for Post Modernity.* Special Education Series. New York, NY: Teachers College Press.

35. Sleeter, C.E. (2008). Preparing White teachers for diverse students. In M. Cochran-Smith, S. Feiman-Nemser, D.J. McIntyre, & K.E. Demers (Eds.), *Handbook of Research on Teacher Education: Enduring Questions in Changing Contexts* (3rd ed.) (pp. 559–582). New York: Routledge.

36. Smith, V.M. (2003). You have to learn who comes with the disability: Students' reflections on service learning experiences with peers labeled with disabilities. *Research and Practice for Persons with Severe Disabilities, 28*(2), 79–90.

37. SOS Healthcare, Inc. (n.d.). Home page. Retrieved from http://scautismhelp.com/

38. Stevens, R. B. J. and Stevens, A. (2005). *Learning in and out of school in diverse environments: Life-long, life-wide, life-deep.* LIFE Center, University of Washington, Stanford University, and SRI International, 2007.

39. Szepkouski, G.M., & Dunn, M.U. (1997). *Qualitative research: A tool to help future special educators see beyond the label of their future students.* Paper presented at the Annual Meeting of the American Association of Colleges for Teacher Education, Phoenix, AZ. (ERIC Document Reproduction Service No. ED406335).

40. Trent, S.C., Kea, C.D., & Oh, K. (2008). Preparing preservice educators for cultural diversity: How far have we come? *Exceptional Children, 74*(3), 328–350.

41. Webb-Johnson, G., Artiles, A.J., Trent, S.C., Jackson, C.W., & Velox, A. (1998). The status of research on multicultural education in teacher education and special education: Problems, pitfalls, and promises. *Remedial and Special Education, 19*(1), 7.

42. Warren, M.R. (2005). Communities and schools: A new view of urban education reform. *Harvard Educational Review, 75*(2), 133–173.

43. Warren, M.R., Hong, S., Rubin, C.H., & Uy, P.S. (2009). Beyond the bake sale: A community-based relational approach to parent engagement in schools. *Teachers College Record, 111* (9), 2209–2254.

44. Welch, M., & James, R.C. (2007). An investigation on the impact of a guided reflection technique in service-learning courses to prepare special educators. *Teacher Education and Special Education: The Journal of the Teacher Education Division of the Council for Exceptional Children, 30*(4), 276–285.

45. Zeichner, K., & Melnick, S. (1996). The role of community field experiences in preparing teachers for cultural diversity. In K. Zeichner, S. Melnick, & M.L. Gomez (Eds.), *Currents of reform in preservice teacher education.* (pp.176–196). New York: Teachers College Press.

CHAPTER 6

THROWBACK TECHNOLOGIES: INTERACTIVE WHITEBOARDS IN SCHOOLS

HEIDI L. SCHNACKENBERG and EDWIN S. VEGA

State University of New York at Plattsburgh, 101 Broad St, Plattsburgh, NY 12901, United States

ABSTRACT

During the past two and half decades, interactive whiteboard technologies have become a mainstay in many schools and businesses. While they may have been cutting-edge and somewhat novel when they were introduced, currently they are outdated educational technology and the monies previously invested in them should be re-appropriated elsewhere. This chapter attempts to explore these issues and offer alternatives to interactive whiteboard technologies in the classroom.

INTRODUCTION

After spending a week preparing a lesson that is heavily dependent on the interactive whiteboard in my classroom, I find out an hour before my for my college-level, Intro to Education class that the board isn't working. Frustrated, but knowing that sort of thing happens, I revise my lesson so that I can teach without the interactive whiteboard. A month later, a brand new board is installed and I'm excited to once again use it during my class. The first morning that I intend to use the board, I go into the classroom early to prep my presentation, video, and other materials. As I try to get the video clip working that I intend to use, I realize that I can't get sound from it. I check the video on my tablet to see if it's the clip itself and find out that it is running fine on that platform. So, I continue to try several ways to get the video to work with sound, but nothing works. Finally, five minutes before my class begins, I give up and call the IT department. To their credit, they send someone down to my classroom right away. However, after fussing around with the board and the video for quite a while, the equally frustrated IT guy tells me, "I have to do more work on this. It's brand new, but I can't figure out what the problem is. It's not anything you did Heidi. The [interactive] whiteboard just isn't working." Although I appreciated knowing that I didn't do anything to break the technology, it was cold comfort given that, once again, I was excited to use it during a lesson and, once again, it wasn't functional.

LITERATURE REVIEW

During the past two and half decades, interactive whiteboard technologies have become a mainstay in many schools and businesses (SMART Technologies, 2015). The boards can access the Internet, display videos and pictures, and play sound bites (Schnackenberg & Heymann, 2013). According to Smaldino, Lowther, & Russell (2012), jpeg images, writing, or entire work sessions can also be recorded and saved for playback later, thus eliminating teachers' need to re-explain work if students have questions. Additional equipment is also available for the boards, such as handheld response systems and portable teacher slates, which enable more student and teacher interactivity (Deubel, 2010). While ever-present in schools and classrooms, current research on interactive whiteboards has yielded mixed results when it comes to student learning gains and/or positive transformations to pedagogy (Higgins, Beauchamp, & Miller, 2007; Smith, Higgins, Wall, & Miller, 2005).

For some teachers, administrators, and students, this technology is the greatest thing to happen to the whiteboards since dry erase markers (Betcher & Lee, 2009; Cruikshank, 2007). Mechling, Gast, and Krupa (2007) found that the boards are effective with whole group instruction on sight words. In a related study, Hall and Higgins (2005) interviewed primary school children about their perceptions of interactive whiteboards and found that the students enjoyed the multimedia capabilities (although they noted associated technical problems) and said that the boards made learning fun. Beauchamp and Parkinson (2005) also recognize the potential of interactive whiteboards for improving science pedagogy by, for instance, displaying interactive models of molecules or frog dissections, and suggest strategies, such as allowing students to manipulate the board to increase active learning for student lessons.

Conversely, Glover, Miller, Averis, and Door (2005) found that most teachers used interactive whiteboards as little more than presentation tools, thereby not changing teaching or learning in any demonstrable ways. Slay, Siebörger, and Hodginson-Williams (2008) stated that although teachers in three South African government schools did value interactive whiteboards, they found them most beneficial when paired with laptop technologies so that student interaction was increased. Wetzel (2010) posits that for the most part, interactive whiteboards were used ineffectively in K-12 science and math lessons due to lack of teacher training and professional development.

In 2009, Salton and Arslan conducted a meta-analysis of research on interactive whiteboards. They found approximately 64 citations for interactive whiteboards in the ERIC database, for literature between 1995 and 2008. Of the 64 citations, only 10 of the publications were experimental research or case studies, while the rest were reports or thought pieces. Therefore, the meta-analysis was conducted on the 10 research studies that were found over the 13-year timeframe. Salton and Arslan (2009) indicate that the findings showed that interactive whiteboards facilitated more professional and efficient delivery of multimedia resources than those teachers typically create and use. They also state that the potential of interactive whiteboards on effective teaching and learning was obvious. Although the power of meta-analysis as a statistical tool is clear (Borenstein & Hedges, 2009; Glass, McGaw, & Smith, 1981), the dearth of research on interactive whiteboards that Salton and Arslan (2009) uncovered makes such bold findings regarding student learning suspect.

While interactive whiteboards may have been cutting edge in the fields of business and industry when they were first introduced in the mid-1990s, and somewhat novel in schools and classrooms when they were marketed there 5–10 years later, in our view they are outdated technologies and the monies previously invested in them should be re-appropriated elsewhere. This chapter attempts to explore these issues and offer alternatives to interactive whiteboard technologies in the classroom.

OUT WITH THE OLD

In a June 2010 article, The Washington Post discussed how educators were questioning the effectiveness of educational technologies, including interactive whiteboards (McCrummen, 2010). Since that time, teachers and even students have continued to seek alternatives to the whiteboards and offer those solutions to broader teaching community via blogs and various forms of social media. The following are a few examples of educators taking this grassroots approach.

In his January 11, 2012, blog post, high-school music teacher and collegiate instructional designer, "Shaun," wonders if Microsoft's newest version of Kinect for Windows will kill the SmartBoard. Kinect tracks one's body movements and voice via a camera and microphone and is primarily for motion-controlled gaming. Clearly, the creation of Kinect did not overthrow the presence of interactive whiteboards in schools, but "Shaun" still makes some excellent points in his post. He describes interactive whiteboards as 1990s technology, where the only motion control and interactivity are via pens (pointers, really) and that only one student at a time can use the board. There are multitouch boards as well, but "Shaun's" point is that gaming technologies have far exceeded the capabilities of the best interactive whiteboards. Currently, kids use virtual reality and motion control regularly when they game, so why not bring those technologies into schools, instead of staying the course with the whiteboards? As James P. Gee (2014) states, the learning involved in video game technologies fits better with the world that today's students live in than do long-established teaching and learning practices that children are exposed to in schools. It is not a stretch then to infer that the teaching and learning practices used in schools are outdated and out of sync with how the rest of the world operates.

Middle school technology coordinator, Benjamin Sheridan, also addresses interactive whiteboards in his blog. In a February 22, 2013, blog post, Sheridan talks about how he is finally parted with the SmartBoard (a popular brand of interactive whiteboard) in his classroom in favor of using AppleTV, both for the multimedia content and the connectivity to individual devices. He cites the SmartBoard as being too teacher-centered since only one person can use it at a time, and that one person is generally the teacher, whereas students are able to create projects on their iPads and share them via AppleTV Airplay.

Perhaps most telling is a 6-second video on Vine posted on September 12, 2014, entitled "Death of a SmartBoard @ms Marshall's" (https:// vine.co/v/OaZWKHulnhJ). In the clip, students in Ms. Marshall's class are laughing as the interactive whiteboard is fairly rapidly flashing various colors, and you hear an adult (very likely the teacher, Ms. Marshall) quip, "It is literally dying right before our eyes." Obviously the event was momentous enough, and funny enough, for CJ Commodore (2014) to post and broadcast it out to the world. But is it really funny, having an interactive whiteboard irreparably malfunction during a class? In 2006, Van Horn suggested that interactive whiteboards were just another technological innovation that won't make much difference for student learning and weren't worth mastering. Perhaps he was right.

Each of the bloggers/vloggers above makes points about the antiquity of interactive whiteboards. Many P-12 students are using gaming technologies of one sort or another, either with powerful home systems, on computers or tablets, or even on smart phones. All of these devices offer more interactivity than an interactive whiteboard and many students use them daily. Students also utilize all of these devices individually, often virtually interfacing with other people (texting, interactive video, photo-sharing, etc.), but they are rarely passive with their technologies. Interactive whiteboards, whether multitouch or single touch, do not offer the opportunity for all students to interact with technology in the way that they typically do when they are not in school. The boards clearly promote a teacher-centered instructional model rather than a student-centered one. Finally, as "Death of a SmartBoard @ms Marshall's" video post indicates interactive whiteboards may have become somewhat of a joke to students in schools. So while they may once have had a motivational/novelty effect on students' learning, currently, that is likely not the case. Interactive whiteboards are old technology and students know it.

IN WITH THE NEW

So if interactive whiteboards are outdated, why are we still installing, repairing, and replacing them in schools? Part of the reason may be the initial monetary investment, or the upkeep of that initial investment. Currently, interactive whiteboards cost several thousand dollars each, for single-user capabilities or multitouch capabilities. However, that money could also be used to put mobile technologies in the hands of students in classrooms so that each student could actively use the technology rather than passively sit and watch interactive whiteboard technologies being used by a teacher (or at most, one or a few other students).

For instance, 20–25 inexpensive laptops, tablets, or even dare we say it, smart phones, would easily equal or be less than the cost of one interactive whiteboard. These individualized technologies, in the hands of the students, would then drive a different instructional model than that of an interactive whiteboard. For the most part, the whiteboards are passive, only allowing from one to a few users at a time (with a multitouch board). Having mobile technologies or individual devices, students could then engage with the lessons and the technologies in a more student-driven and interactive approach to learning. Students could work separately or in small groups to solve problems, engage in situated learning experiences, think critically about how to resolve scenarios, and/or investigate questions about which they are curious. Of course they can do all of those things without mobile or individualized technical devices, but the ability to find out information or reach experts without these technologies is much more difficult.

Another alternative to investing in interactive whiteboards would be to equip schools with current, interactive, learning games. These can be loaded onto laptops, tablets, or even engaged through specific gaming systems. At home, many students utilize these types of gaming technologies regularly. The interfaces include virtual reality, social media, simulated real-life problems, team building, cooperation, research, and perhaps video and photo sharing. These are the types of technologies that students are familiar with, find fun and enjoy, and would be more motivated to engage with in a school setting, rather than sitting and watching a teacher operate an interactive whiteboard.

While students may well be happy, or at least ambivalent, about the idea of not having an interactive whiteboard in their classrooms, teachers will be hard pressed to give up the option to project web interfaces, student work, or to simply "have things up on the board" in the front of the room. Teachers can easily have these capabilities without the added cost and usability issues of interactive whiteboards. At a minimum, teachers could simply use a computer and projection system to show most of what they need to their student body. If they want to project student's work, using software that connects to each student's laptop or handheld device to a teaching station is currently commonplace in many schools and would serve that purpose well. A simple document camera and projection system also display materials and student work for analysis very nicely, at a fraction of the cost of a whiteboard. Or, perhaps even a system like AppleTV in conjunction with individual devices would be a better investment than an interactive or multitouch whiteboard. And while the whiteboards required a significant amount of professional development in order to learn to operate them effectively, the technologies suggested here are either quite simple to use or devices that both teachers and students use regularly outside of school, so very little professional development would likely be necessary.

CONCLUSION

For the thousands of dollars that have been invested in interactive whiteboards in classrooms, there are many other, current, technological options that will better serve students and schools in the long run. Generally, after 20–25 years, even with upgrades, technologies are considered obsolete and are replaced by more current hardware and software that allow users to do more complex tasks more easily than with the old equipment. It is time for administrators, teachers, and school districts to make a change and stop investing, or reinvesting, in old technologies that no longer hold students' interest and have never clearly been proven to improve students' learning. The variety of technologies that could furnish a classroom, in place of interactive whiteboards, is vast. It is up to the various constituents in a school to decide what types of devices best suit the learning needs of their students. Regardless of the option, or various options, that they choose, it is time to let the phase of interactive whiteboards in schools move quietly into the past, and an age of more interactive, mobile technologies in schools takes their place.

KEYWORDS

- **education**
- **learning**
- **SmartBoard**
- **students**
- **teaching**
- **technology**

REFERENCES

1. Beauchamp, G. & Parkinson, J. (2005). Beyond the 'wow' factor: Developing interactivity with the interactive whiteboard. *School Science Review, 86(316)*, 97–103.
2. Betcher, C. & Lee, M. (2009). *The interactive whiteboard revolution.* London, England: ACER Press.
3. Borenstein, M. & Hedges, L.V. (2009). *Introduction to meta-analysis.* Hoboken, NJ: John Wiley & Sons, Inc.
4. Commodore, C.J. (September 12, 2014). *Death of a SmartBoard @ms Marshall's* [Video file]. Retrieved from https://vine.co/v/OaZWKHulnhJ.
5. Cruikshank, D. (October, 2007). *Board of education: A wall-mounted computer monitor for your classroom.* Retrieved June 10, 2016 at http://www.edutopia.org/interactive-whiteboards
6. Deubel, P. (August 4, 2010). Interactive whiteboards: Truths and consequences. *THE Journal.* Retrieved February 8, 2011, from http://thejournal.com/Articles/2010/08/04/Interactive-Whiteboards-Truths-and-Consequences.aspx?Page=1.
7. Gee, J.P. (2014). *What video games have to teach us about learning and literacy.* Hampshire, England: Macmillan.
8. Glass, G.V., McGaw, B., & Smith, M.L. (1981). Meta-analysis in social research. *Sage Library of Social Research*, 124.
9. Glover, D., Miller, D., Averis, D., & Door, V. (2005). The interactive whiteboard: A literature survey. *Technology, Pedagogy and Education*, 14(2), 155–170.
10. Hall, I. & Higgins, S. (2005). Primary school students' perceptions of interactive whiteboards. *Journal of Computer Assisted Learning*, 21(2), 102–117.
11. Higgins, S., Beauchamp, G., & Miller, D. (2007). Reviewing the literature on interactive whiteboards. *Learning, Media, and Technology*, 32(3), 213–225.

12. McCrummen, S. (June 11, 2010). Some educators question if whiteboards, other high-tech tools raise achievement. *The Washington Post.*

13. Mechling, L.C., Gast, D.L., & Krupa, K. (2007). Impact of SMART board technology: An investigation of sight word reading and observational learning. *Journal of Autism Development Disorder, 37,* 1869–1882.

14. Salton, F. & Arslan, K. (2009). A new teacher tool, interactive white boards: A meta analysis. In Gibson, R., Weber, K., McFerrin, R., Carlsen, R., & Willis, D. (Eds.), *Proceedings of the Society for Information Technology & Teacher Education International Conference 2009* (pp. 2115–2120). Chesapeake, VA: AACE.

15. Schnackenberg, H.L. & Heymann, M.J. (2013). Interactive whiteboards: Worth the investment? A case study at Hawkins Elementary School. *Journal of Cases on Information Technology, 14(1),* 15–25.

16. Shaun (2012, January 11). *Did Microsoft just kill the SmartBoard?* [Web log post]. Retrieved from http://hollandsopus.wordpress.com/2012/01/11/did-microsoft-just-kill-the-smartboard/.

17. Sheridan, B. (2013, February 22). *Goodbye SmartBoard..... Hello AppleTV* [Web log post]. Retrieved from http://www.coetail.com/bsheridan/2013/02/22/goodbye-smartboard-hello-apple-tv/.

18. Slay, H., Siebörger, I., & Hodginson-Williams, C. (2008). Interactive whiteboards: Real beauty or just "lipstick"? *Computers & Education,* 51(3), 1321–1341.

19. Smaldino, S.E., Lowther, D.L., & Russell, J.D. (2012). *Instructional technology and media for learning* (10th ed.). Boston, MA: Pearson Education Incorporated.

20. SMART Technologies (2015). *20+ years of innovation.* Retrieved June 22, 2015, from http://smarttech.com/us/About+SMART/About+SMART/Innovation/20+years+of+innovation.

21. Smith, H., Higgins, S., Wall, K., & Miller, J. (2005). Interactive whiteboards: Boon or bandwagon? A critical review of the literature. *Journal of Computer Assisted Learning,* 21(2), 91–101.

22. van Horn, R. (2006). The technology penalty. *The Phi Delta Kappan,* 87(9), 647–709.

23. Wetzel, D. (July 28, 2010). Why interactive white boards are used ineffectively in classrooms. *Teach Science and Math.* Retrieved February 6, 2011, from http://www.teachscienceandmath.com/2010/07/28/why-interactive-white-boards-are-used-ineffectively-in-classrooms.

CHAPTER 7

AT YOUR SERVICE: A COLLEGE AT WORK WITH ITS COMMUNITY AS A MEANS OF OUTREACH AND MUTUAL ENRICHMENT

KERRI ZAPPALA-PIEMME and DAVID IASEVOLI

State University of New York at Plattsburgh, 101 Broad St, Plattsburgh, NY 12901, United States

ABSTRACT

What constitutes "service" in the academy? In this study, the authors review the classic literature in the field and conduct mixed-methods research to uncover the values promoted by their college. They participated in the creation of a conference about "diversity" in Upstate New York, and tracked the impact of the conference upon participants. Through surveys and questionnaires, interviews, and "office conversations," they illustrate several findings. First, the requirement of "service" for tenure-track faculty remains both flexible and amorphous. Second, subjects who work for the college note that much in-house service—such as that which is required by committee-work—often feels solipsistic. Finally, they and the authors advocate for more service that is explicitly conducted for the public good by reaching out beyond the immediate campus.

BACKGROUND

In the Fall 2013 issue of *Thought & Action* (Golden, 2013), Robert Golden, the former Provost and Branch Campus Dean for State

University of New York (SUNY) Plattsburgh, created a portrait of a dystopian university system 50 years hence. He imagined a virtual campus administered by a "senior branch manager" instead of a college president, students progressing through "statewide Learning Outcomes Metrics Analysis (LOMAs)," and college support networks outsourced to Nigeria—"the current low cost port-of-call for firms seeking to reduce costs to the bare minimum" (p. 47). When he visited our campus to discuss his article, Golden noted that if current trends continue and our colleges increasingly follow corporate models, then his vision for the public university may become a reality much earlier than 2050.

With this jeremiad in mind, we wanted to explore the possibilities of a small, rural[1] branch campus' efforts to reach out into the immediate community to "make a difference": a mutual enrichment of our college and its external community was the central goal. Golden's essay not only concerns the erosion of "professionalism" on college campuses but also strikes at the heart of any college's *significance* in this era in which the market economy supersedes all other matters. Our monthly staff-faculty meetings always include statistics about the increase or decrease of enrollment, program by program. We discuss the possible benefits of "branding" on the part of our college, and ways to enhance its profile in New York's capital region and to compete against other local colleges. In short, the majority of meetings revolve around business as usual. Yet a sense of *purpose* that means much more than the bottom line lingers on the public college campus. As a public institution, what do we *purport* to effect? "Public schools are not merely schools for the public ... but schools of publicness; institutions where we learn what it means to be a public" (Barber, 1992, p. 222). On our small SUNY Branch Campus, we continue to address the possibilities of *authentic service* to the general public.

Other than offer opportunities for bachelor's degrees, what services can and should a college campus offer to the community in which it is

---

[1]The term "rural" can be problematic, as one primary definition derives from its contrast to "urban": "Whatever is not urban is considered 'rural'" by the U.S. Census Bureau (U.S. Department of Health and Human Services, Health Resources and Service Administration, 2014). Thus, for our purposes here, we use the term to signify that even though we do not teach in an agricultural community, its population numbers less than 50,000.

located? Furthermore, how can we serve something of value to a rural community? Finally, in what ways can we insure that we are offering programs and services that are consistent with research-based and best practices? These were some of the questions that we considered both at the start of our plans for a conference and as we undertook this research. What follows is a brief recapitulation of the steps we took to realize this conference.

In September 2013, a small group of college faculty and several students formed a planning committee to organize an interdisciplinary conference for the greater community of Queensbury-Glens Falls, NY. We represented the criminal justice, business, school leadership, and education programs headquartered on our main campus in Plattsburgh, NY. Our primary goal was to provide information and resources about such topics as poverty, law enforcement, racial tensions, and higher education to any who wished to attend; there was no registration fee. Several members debated this last condition, but we finally agreed that a "free conference" stayed truer to the desire to perform *pro bono* service.

Beneath the rubric of "North Country Diversity Conference," we petitioned our main campus for funding and received $7500 from the Grant and Sally Webb Endowment Fund through Plattsburgh College. The mission of this endowment fund is to "support lectures as well as other events that overcome differences and build a common foundation in multicultural awareness." We sent out a "Call for Participation" (Appendix A) to North Country schools and businesses, government agencies, and public service organizations, with such questions as follows:

- What changes in demographics can we expect in the North Country?
- What effects will immigration have upon small communities?
- Is integration dead?
- How will public schools accommodate multiculturalism?
- What type of leadership is needed in the North Country?
- How are service agencies adjusting to the changing landscape?
- What are the culture-related best-practices or issues we face?
- *What questions about changes in our community do you wish to address?*

We planned our conference to begin on 19 June—"Juneteenth"—to acknowledge the anniversary of the date in 1865 when many Blacks in the South, former slaves, first learned of the 1862 Emancipation Proclamation. We sent out letters of invitation to such luminaries as the nation's First Lady and the Governor of New York State. (Both had previous engagements. The Governor's office delegated Deputy Deirdre Scozzafava to attend, and she delivered our keynote address.) Faculty from our main campus were invited and encouraged to submit proposals for their workshops; thus, we created a peer-review process for junior faculty who were interested. Our business professor regularly edited and updated the website for the conference. Between October and May, we created plans for a 2-day event that attracted 112 participants and that included music, workshops and lectures, a panel discussion, and hot-air balloons (but no flights) (Appendix B—brochure). Throughout our meetings, we regularly tested ourselves: How does this event serve the local community? Yet we never looked at the historical significance of college service during these organizational meetings.

REVIEW OF LITERATURE

Faculty service is inconsistently defined in higher education research (Ward, 2003). "Internal service" supports the institution's mission, operations, and cultural life (e.g., service on a curriculum committee or graduation), whereas external service involves outreach and public service (e.g., as a business professor works with the town of Queensbury's Rotary Club), or support to disciplines or fields and professional associations (e.g., the author works on the review panel for his union's journal of higher education). Scholarly service draws on subject matters in which professors claim substantive expertise (e.g., a professor of education works with a local school to train "teacher leaders"); non-scholarly service is disconnected from a professors' subject-matter expertise (e.g., the author volunteers on his town's Board of Ethics).

Whatever "faculty service" should mean, its significance depends upon its *place*. The contingent quality of internal service further complicates such differences in mission that are characteristic of institutional type. Service is common to all types of nonprofit post-secondary educational institutions— for example, community colleges, private liberal arts colleges, and public and private universities. Faculty service is broad-ranging, eclectic, and always contextualized. Given this internal diversity, it is no surprise that service is underdefined (Neumann and LaPointe Terosky, 2007, pp. 282–283). Nevertheless, these flexible concepts of service are based on the specialized knowledge acquired by faculty and administrators in the colleges and universities through a variety of activities. Various scholars have identified different elements of service activities. Harper and Davidson (1981) suggested the following types: dissemination of knowledge beyond the campus, delivery of instructional programs beyond the campus, using applied research for immediate application by the public, sharing of resources, and the development of public policy issues and alternatives. In their classic study, Henderson and Henderson (1975) expand upon these ideas by including pure and applied research along with "extension services and continuing education, community services, consulting services, reference bureaus, and cultural events." (p. 112)

These historical considerations of college service reverberate in current definitions. Neumann and LaPointe Terosky (2007) state that "Professors create their careers through three forms of work: research, teaching, and service. Teaching and research are well-defined in most professors' careers and in higher education at large. However, faculty service is nebulous" (p. 282). Abukari (2010) agrees: "In the field of higher education, the service concept can be ambiguous and contestable" (p. 44). She continues, in her two case studies, to illustrate that university or college service is a "broad concept" that covers such "core activities" as outreach, community service, and service learning. And it must always be contextualized: for example, in a research university, faculty service on a committee goes farther when such committee work intersects with an instructor's area of expertise. Our college, for contrast, is more likely to acknowledge service that "reaches out."

We have found one definition of college service that resonates best with us: *work that develops knowledge and draws upon professional expertise for the welfare of society* (Checkoway, 2001, p. 143, emphasis added). This returns us to a historical perspective on colleges, and what may be, to many, the *raison d'etre* for post-secondary institutions: teaching and research have always been and remain today a form of service.

In general, service is most often cited as one of the main criteria used for promotion or for tenure in universities; yet the reality is that college service can weigh relatively little. In the triumvirate of promotion criteria, service becomes a lady-in-waiting to teaching and scholarship. It simply did not and still does not rank as highly as the latter two strands for promotion. There are three reasons for this. First, as mentioned above, service is hard to define and there are no common metrics to use to differentiate between professionally related and non-professionally related service (Florestano and Hambrick, 1984). Even in the latest research, a provost asked a college-wide committee to develop criteria to assess service, but did not mandate that the criteria be used to evaluate faculty for promotion and tenure (O'Meara, 2002, p. 65). Simply (and perhaps crassly) put, there are no good measures to evaluate service. Second, faculty and administrators generally do not regard service highly. Some educators do not believe that service is a legitimate university function and that there is little equality of service among institutions or within institutions. The academic culture traditionally holds research, publishing, and notoriety at the zenith of its profession (Antonio, Astin, & Cress, 2000, p. 388), and professional services and teaching are often given less weight than research in tenure reviews (O'Meara, 2002). Third, many college administrations tend to reward faculty who *receive* the kind of service that is typically most valued within the institution—grants and awards. In other words, the value of a faculty member is often commensurate with the grant monies she brings *into* the college: as the award of external grants enhances the institution's reputation, the recipient "does her campus a service."

We found remarkably little in our university's official description of service. In the agreement between the SUNY College of Arts and Science and our union, there is one sentence: "Effectiveness of University Service—as demonstrated by such things as college, University, public service, committee work, administrative work and work with students or community in addition to formal teacher-student relationships" (p. 11–12). There is also a longer list of "elaborations" that suggests activities that may (or may not) be expected of faculty in SUNY Plattsburgh's Education Unit (2008, p. 3):

- Program Area Coordinator
- Program Leader
- Unit and/or College Committee Chair
- Unit and/or College Committee Member
- Clinic Director
- Faculty Senator

- Faculty Advisor for Student Organization
- Coordinator of Special Departmental Services
- Professional Consultant (ongoing) with a school, agency, institution, organization, etc.
- Guest Speaker
- Member of a state, national, and/or international professional organization
- Executive member of a state, national, and/or international professional organization

Note that the closest these activities get to work with local groups—outside of the college—is the item "Professional Consultant."

Now, even though consulting for a group outside of the college can fall under the loose rubric of "service," this is typically for financial gain, and we are concerned here with the work that college faculty members may perform *pro bono*—for the "public good." Again, this notion gets back to the historical roots of post-secondary education in the United States. Thomas Jefferson wrote to James Madison, in 1787, "The only sure reliance for the preservation of our liberty is to educate and inform the whole mass of the people" (Barber, 1992, p. 224). How can a college address the education of "the whole mass of the people"? As Lustig (2011) argues, our system of colleges and universities, both private and state-funded, *depends upon* the greater community to maintain what he labels "the knowledge commons … a place in which the cultural and intellectual wealth of the past is made available, where ideas are freely shared and where ideas also grow by cross-fertilization from many fields" (p. 15). Has this notion of a college's need to contribute to the public good, in order for the college itself to flourish, now become antiquated, or even arcane? We, the authors, think otherwise: the contemporary college campus cannot afford to seem "obscure" and secret to the general public, but, rather, needs to progress toward a more *exoteric* presence. With this provision in mind, we planned a conference to serve the greater community—outside of the college.

METHODOLOGY

Immediately after the conference, informal feedback gave us the impression that it was an unqualified success. Unfortunately, only 33% of the participants completed a nine-item evaluation survey (Appendix C). Therefore, we, the authors, decided to attempt a more thorough analysis of the proceedings. We reached out to some 20 participants—students and faculty and presenters—and interviewed them. Most of the interviews took place on either of Plattsburgh's campuses, and each lasted between 8 and 20 minutes. (Henceforth, we refer to these 20 as "interview subjects" to distinguish them from the universal set of "participants" who attended the conference.) We met during September and October 2014. We guided our talks with the following four questions:

1. What else should a small college campus provide for its community?
2. What is the college's most valuable resource?
3. What should a conference workshop *do*—teach its participants to do something new?
4. What does college mean now that there are no jobs?

But these questions served merely as signposts, and the interviews were open-ended: we asked subjects to comment on anything about the conference that struck them as noteworthy.

Now, a common mission among academic institutions is to educate and meet the unique needs of every student in a safe environment, to provide opportunities that expand interests, enhance abilities, and provide all students with knowledge, skills, and character essential to being responsible citizens in a diverse society. We entered into our dialogues with interview subjects with the following assumptions: the value of a bachelor's degree has changed and shall continue to change; college campuses can still function as loci for the vital exchanges of ideas; faculty have a duty to serve their college's greater community; and, most significantly here, *external service—to the greater college community—matters more than internal service, since it lies close to the central purpose of higher education.* Such assumptions may have shaped the responses from interview subjects, inasmuch as they realized that we were determined to create stronger experiences for the college to serve the general public.

Furthermore, we recognized immediately the following limitations and delimitations: a very small sample is involved; this is a rural college campus and any findings or implications may not resonate at all within either large universities or urban settings. Finally, we proceeded with the following irony uppermost in mind: there are no faculty members of color on this branch campus, and, further, there are few students of color—in short, our immediate college community is *not racially diverse*. Nevertheless, we launched our inquiries in order to plan for our next conference—that draws in a greater variety of participants and presenters—and to create a platform to discuss and change our lack of diversity within an increasingly diverse nation.

FINDINGS

Our interview subjects corroborated what the initial surveys suggested, and, as their descriptions elaborated upon these rudimentary data, we detected three overarching themes for their responses: Creation of Networks, an Expanded Definition of College, and Redefinitions of College Service.

CREATION OF NETWORKS

We predicted that many of the participants in the conference and the respondents to both our survey and interviews would note the centrality of *networking*. The first question on the survey was: How did you hear or learn about this conference? Forty percent of the participants indicated that they learned about the event from a colleague or a friend, and 26% learned about it through e-mail or newsletter. Therefore, the most effective mode of disseminating information about the conference was through word of mouth and networking (Appendix D, Graph, and Table 1). Networking, of course, is intrinsic to any conference. As our Branch Campus Dean noted, "ultimately, it's the place to develop networking—isn't networking ultimately it?" But he stretched this dynamic, when he emphasized *the differences* that were noteworthy within our particular conference: "... a small campus provides a forum for exchanging ideas, for *bringing disparate groups together*" (emphasis added). When participants were asked to specify the main reason for attending the conference, 7% indicated that they attended the conference to network (Appendix D, Graph, and Table 2). Concordantly, community organizations, such as Family YMCA of the Glens Falls Area, the Warren-Washington Association for Mental

Health, Inc. (WWAMH), and the Community, Work & Independence, Inc. (CWI), set up booths throughout the conference to network and provided participants with information regarding their services. A participant specifically stated that through this venue she/he "discovered new resources and potential collaborations."

Several interview subjects noted the symbiotic relation that exists between a small college and its surrounding community. The professor who served as the main facilitator for the conference pointed out that students automatically become a part of a community greater than that of the college—"our students do not live just on a college campus, they live in the community." He emphasized our college's responsibility to make this situation more overt and to "build those connections ... and relationships." Another subject pointed to an "organic" relationship that already exists between the college and those non-academics who may have a vested interest in a particular topic—such as the apparent lack of racial diversity in rural Upstate New York. Throughout the surveys, many noted that participants benefited from meeting other professionals in the community who hold similar interests and with whom they could network. Finally, the networking dynamic, according to one subject, creates a cycle: "Networks ... allow you to be competitive ... building a certain level of social capital that is going to pay off for you." In other words, the more that we, as faculty members of a small college, create programs that serve the world outside the campus, the more we can expect such entities as local businesses to value our presence as a resource that contributes to their growth.

AN EXPANDED DEFINITION OF COLLEGE

A majority of respondents commented on the facticity of a small college as a kind of hybrid entity—one that no longer stands only or even primarily as a place for acquiring a degree. In other words, the tradition of the "Ivory Tower," viz. an institution not merely divorced from mundane matters but placed *above* such concerns, has receded into a historical background, where we think it belongs. Our Branch Campus Dean described our central purpose as "all about serving the community ... Make this a hub."

A psychology professor echoed this description when he said that he hoped for our college to "provide ... an intellectual beacon for the town." His image referred to the potential ability of our campus to bring in more members of the general community. He also emphasized that our nonprofit profile created opportunities to share a kind of "un-biased knowledge" with the community, that is, since the college bears no official affiliations with any for-profit institutions, the kind of information that we can provide is, in effect, "duty-free."

There were 30 individuals, from both the college and its greater community, who shared their knowledge during the conference through presentations, as speakers and panel members. The evaluation survey asked participants which individuals they were most interested in listening to, and they were able to select more than one person. Participants indicated that they were most interested in listening to faculty from local colleges (33%), community leaders that serve in government agencies (28%), public service directors of local and national community organizations (13%), local businesses (13%), and chief school officers (9%) (Appendix D, Graph, and Table 3). A participant noted on the survey that, "Bringing together a significant number of leaders in our regions provides us with a foundation to address our diversity issues with some critical mass."

Furthermore, a good number of interview subjects pointed to the importance of the college as a "bridge":

Maybe schools, universities, can be more a bridge between employers and university itself.
[Our conference] can be preparation for creating employment opportunities.
Relationships, partnerships, [and] joining people in collaborative efforts ... lead to more opportunities for careers.

The resident professor of business on our branch campus described the critical importance of the college putting itself at the "disposal" of its greater community, because "how much we consider our resources to be hard to replace [to the local population] then translates into being valuable."

Several respondents highlighted a transformative value of the college to the surrounding community:

Faculty ... help prepare students as future leaders and decision makers in our communities.

[The] value of education is greater than the degree or skill set that is learned.

[We try] to provide the "enriching" aspects of education and give students experiences that are not just classroom related.

All of the interview subjects expressed their hopes—and excitement—that this process of, say, an "extra-collegial" engagement had already reaped dividends, as when one professor noted "when the students leave, they really never left." He also pointed to the importance of a dialectical relationship between academics and members of the community, when he said that "everyone has an opportunity to teach everyone else." In their ongoing dialogues, instructors, students, and non-academics create a cycle of teaching and learning—professors become students, even as our county sheriff became an instructor, and then a student himself, for example. "Teaching and learning will probably be bi-directional," noted another professor, about the possibility of future conferences here. When we pushed this subject to consider the relevance of college itself in our current situation where the bottom-line governs all, she said "If college is a process of learning about the world as well as self, in order to live a better life, and/or make life better for others, then *it means a lot*" [original emphasis].

SERVICE REDEFINITIONS

We heard relatively few participants and interview subjects actually mention specific recommendations for services that the college could and perhaps would do better to provide, but these included such wish-list items as a job fair, cultural events, business partnerships, a campus child-care center and/or preschool, a health and wellness center, a career management center, and an intercultural meeting space. We, the authors, see great value in the creation of at least some of these centers and events, such as the community's use of the library and computer centers. And we shall engage our administration in dialogues about their feasibility (viz. costs).

We also noticed that several interview subjects wished to emphasize that the community serves the college in a variety of ways:

The community also has something to bring, members of the community too.
[The community] has endless resources to share with the college.
It isn't just about academics ... the general population [here] has much larger concerns that impact the college.

Participants wrote on the survey that the most beneficial aspect of the conference was the ongoing discussions that took place between audience, panel members, and speakers—not only in workshops but also in the hallways and especially during lunch. In the final event, everyone was present in a lecture hall—panel members, lead facilitator, workshop leaders, and general audience members, and some noted that the "engaging audience" added to the conference; they had an "opportunity to share their own experiences" and these "personal interactions will forever be remembered."

Thus, we begin to see that the college's attempts to reach out to serve the greater community actually achieved the start of a *unification of efforts* on the part of college faculty, students, and community programs. Furthermore, participants suggested that a conference should explicitly "reach out ... shar[e] resources," and that "faculty members ... serve as role models and mentors for students in regard to community membership and contributions." Another academic noted that one of the intrinsic purposes of any college is to "center" students in their own community-fellowship: "In college, students ... can identify interests and passions *they didn't even know they had—and experience what it means to live in and contribute to a community."* And, as a necessary consequence, this "helps us to move from theoretical ideas to possible action." In short, we heard comments about *praxis*—the hybrid of theory and practice that we so often search for on a college campus. As Freire (1993) so strongly enjoins us,

Liberation is a praxis: the action of men and women upon their world, in order to transform it ... Education as the practice of freedom...denies that the world exists as a reality apart from the people ... human activity consists of action and reflection: it is praxis; it is transformation of the world. (pp. 60 & 62)

Finally, we noted a healthy tension around the very notion of "service" itself as it plays on our college campus. We heard one instructor (who attended but did not present at the conference) note that "The College needs faculty to serve *within*—[this is] more important than service to the community." Another countered that a conference for the greater community "should count *significantly* as part of the service component to the college." Furthermore, he argued that many college committees often do not accomplish much—"a lot of it is mindless." With such responses in mind, we formed the following recommendations.

FURTHER IMPLICATIONS

The original title for our conference in June 2014 was "The More Things Change." We imagined that some participants would engage with the irony of our region's *apparent* lack of diversity and perhaps even argue that nothing much has changed in the past 20 years—the period in which our college has run a branch campus here—or that we have to work harder to effect real change. In a similar vein, we, the authors, realize that "the more you do for your college, the more you need to do." In other words, these experiences of planning, enacting, and researching a small conference have taught us that we must "up the ante" to create our next conference that shall serve greater numbers.

First, we need to create new ways to attract more participants. In order to attract the greatest number of participants, we need to address the questions we posed at the beginning of this essay: what services can and should a college campus offer to the community in *which it is located? Further, how can we serve something of value to a rural community? Finally, in what ways can we insure that we are offering programs and services that are consistent with research-based and best practices?* In response to the last two survey items regarding future conferences, 97% stated they would attend next year's diversity conference (Appendix D, Graph, and Table 4), and the topics and presenters they would like to hear are as follows:

- I would like to learn how educators can teach our children a language of diversity.
- Diversity training tangible work being done.
- More on programs and strategies for elevating the socio-economic disadvantaged.

- Topics: Poverty, Economic diversity and language differences.
- Collaborative efforts from different racial groups working together to address poverty as a form of diversity that crosses racial boundaries.
- I would like to hear from either a Plattsburgh St. College or SUNY Albany H.R. individual as to their policy in hiring minorities for their campuses and the effect this multi-cultural approach has had for the campus.
- I think there should be topics on LGBTQ issues, religion, ableism, classism, race. I was a little disappointed at some of the mini workshops. I didn't think there was a lot of focus on diversity in the North Country or how to offer appropriate service provision to the "diverse" population. I felt there were too many statistics and not enough experiential components. It was a great first try at the conference but I just wish there was more information on how to work with diverse individuals that are living in the North Country which historically has not been a diverse area.

Second, we need to create regular opportunities for community members to join college faculty for brainstorming sessions, cultural events, lectures, etc. This becomes a kind of hermeneutic circle of significance: the college grows in importance to the lives of others even as those immediately outside it change it for the better—by taking advantage of the college's resources. One of the psychology professors on our branch campus put this best, in a follow-up interview:

When community members take advantage of college resources we form a bond with the community ... a symbiosis where new and pathbreaking achievements can be made. [It] not only provides our students with new opportunities to apply the knowledge and skills the college gives them, but to give back to their community.

How often do faculty consider the fact that we and our institutions do not exist without the support of the community? Recently, attacks upon the usefulness of "the college experience" (and especially the liberal arts education) have intensified, and this is due, at least in part, to the public perception that college is hermetic—an entity unto itself. We think that the conceptions of "faculty service" need to become more "other-centric" in order for the college to claim—or reclaim—a position of significance. We have found that a college tends to *keep service to itself* all too often: internal service, such as committee work, is indeed necessary for the college's very sense of definition. Yet it is not enough. College service needs to keep extending its faculty to serve others—those outside of the tight circle of the Academe. We think that when our university, the SUNY system, explicitly shows faculty, especially "junior faculty," that such service counts far more than service on an internal committee, say, then its value to the general population increases significantly.

Finally, we suggest that our college must continue to search for and create means to work for the "public good," because when it keeps service to itself, and when faculty and administration are preoccupied with the business aspects of higher education, then we retreat to the Ivory Tower—and incur negative perceptions from the general population. As Golden (2013) concludes his essay, "As community bonds have weakened and common understandings about the value of education or obligations to others have deteriorated, a narrow set of quantitative measures have rushed in to fill the void" (p. 51). We suggest that this is a pressing opportunity for our college to take a few giant steps away from such a business of quantification.

As our Spring semester draws to a close, we can already highlight certain measures that bring us closer to this ideal of service for the public good. At the start of the year, our Dean organized and hosted the first "Climate Reality Conference" on our branch campus. This event drew in a hundred participants who were, for the most part, not academics. Local businesses, such as a solar installation company and Scientific Environmental Activities Watershed Alliance Sound Partnerships (SEA WASP), provided resources. The Dean promises that this shall become an annual event. Furthermore, he has begun a concerted effort to visit every high school in the Greater Capitol Region to provide information about our college and its programs. Job fairs have occurred in our local high schools. Finally, we have plans for another conference, centered on poverty, for March 2016. We have decided to elicit prospective participants on two fronts: first, we shall invite fellow college instructors to present; second, we shall call for area residents to tell their stories and to lead us into areas where we learn more about the best ways to serve. We aim to make sure that this is not another conference that merely serves the academic institution, but, rather, one that actually delivers the goods to those whose needs do not resemble a college's typical needs.

In a more perfect—and, in a way, a more pragmatic—world, our college campus becomes a regular center for community gatherings. We resurrect the classical Greek site of the *lyceum* or the Roman *forum*, a place for "public-ness" for an at least monthly engagement with ideas, problems, arguments, and mutual efforts toward solutions—the college as a locus for truly authentic service. We think that Nobelist Toni Morrison (2001), in an address to her own home institution, Princeton, puts this best:

> *If the university does not take seriously and rigorously its role as guardian of wider civic freedoms, as interrogator of more and more complex ethical problems, as servant and preserver of deeper democratic practices, then some other regime or ménage of regimes will do it for us, in spite of us, and without us (p. 278).*

KEYWORDS

- **Community, Work & Independence, Inc.**
- **Learning Outcomes Metrics Analysis**
- ***pro bono* service**
- **service**
- **Warren-Washington Association for Mental Health, Inc.**

REFERENCES

1. Abukari, A. (2010). The dynamics of service of higher education: A comparative study. Compare: A Journal of Comparative and International Education, 40(*1*), 43–57.
2. Antonio, A.L., Astin, H.S. & Cress, C.M. (2000). Community service in higher education: A look at the Nation's faculty. The Reviewer of Higher Education, 23(*4*), 373–397.
3. Barber, B. (1992). *An aristocracy of everyone: The politics of education and future of America.* New York: Ballantine.
4. Checkoway, B. (2001). Renewing the civic mission of the American research university. *Journal of Higher Education*, 72(2), 125–147.
5. Florestano, P.S. & Hambrick, R. (1984). Rewarding faculty members for profession-related public service. *Educational Record*, 65(*1*), 18–21.

6. Freire, P. *Pedagogy of the oppressed* (Ramos, M.B., trans.). (1993). New York: Continuum.
7. Golden, R. (2013). Northern twilight: SUNY and the decline of the public comprehensive college. *Thought & Action*, 29, 45–56.
8. Harper, C.L. & Davidson, C. (1981, July). Faculty public service. *Bovius*, 1(*3*), 5–11.
9. Henderson, A.D. & Henderson, J.G. (1975). *Higher education in America: Problems, priorities, and prospects*. San Francisco: Jossey-Bass.
10. Lustig, J. (2011). The university besieged. *Thought & Action*, 27, 7–22.
11. Morrison, T. (2001). How Can Values Be Taught in This University? *Michigan Quarterly Review*, 278.
12. Neumann, A. & LaPointe Terosky, A. (2007). To give and to receive: Recently tenured professors' experiences of service in major research universities. *Journal of Higher Education*, 78(*3*), 282–310.
13. O'Meara, K.A. (2002). Uncovering the values in faculty evaluation of service as scholarship. *The Review of Higher Education*, 26(*1*), 57–80.
14. Performance Reviews of Academic Employees: Policies and Procedures. (2005). *An agreement between State University of New York College of Arts and Science at Plattsburgh and United University Professions* (Third Edition).
15. "SUNY Plattsburgh Education Unit, Tenure and Promotion Elaborations" (Draft 2008).
16. Ward, K. (2003). *Faculty service roles and the scholarship of engagement: ASHE-ERIC higher education report*. San Francisco: Jossey-Bass.

CHAPTER 8

RESISTANCE TO RACISM IN THE SOCIAL STUDIES: REFLECTIONS, PEDAGOGIES, AND CRITICALITY

EMILY A. DANIELS

State University of New York at Plattsburgh, 101 Broad St, Plattsburgh, NY 12901, United States

ABSTRACT

For educators interested in the impact of racism on the curriculum, students, and schools, the social studies provide a fascinating arena for analysis because of their broad range of subjects and connection to citizenship and "democracy." The social studies are extremely multifaceted; they are also amorphous and constantly struggling for an identity as a discipline (Jorgensen, 2014; Lee, 2005). The theme which guides this review of the literature focuses specifically on active resistance to racism within the social studies. More specifically, hopeful and critical practices are the focus of this review. By using the terms hope and criticality in conjunction, I wish to focus on the possibilities of highlighting practices that involve both of these standpoints and experiences. Criticality is important to highlight and grapple with persistently cruel racial injustices, as well as potentially transformative approaches that do not simply recreate the status quo, while hope is important for the continuation of such efforts and for the empowerment of those involved in this challenging arena. By speaking of the active agency of teachers and students in resistance, I would like to highlight the crucial nature of synergistic projects that involve not just teacher-directed changes, but those that originate from the students themselves as well.

INTRODUCTION

"We are what we know." We are, however, also what we do not know. If what we know about ourselves our history, our culture, our national identity—is deformed by absences, denials and incompleteness, then our identity—both as individuals and as Americans—is fragmented. (Pinar, 1993, p. 61)

BACKGROUND: THE SOCIAL STUDIES AND RESISTING RACISM

The social studies are simultaneously a vital arena and a problematic one within which to explore themes of racism and resistance. They have shifted and changed definitions and purposes, with a variety of focuses depending on the particular era and political and social climates. Currently there are still debates among educators as to the purpose and goals within social studies, such as which areas must be focused on, whether it is history, economics, civics, or citizenship education, especially in the era of high stakes testing. As argued by Ross, Mathison, and Vinson (2014), "it is not surprising that social studies has been racked by intellectual battles over its purpose, content and pedagogy since its very inception as a school subject in the early part of the 20th century" (p. 25), due to the many perceptions, misperceptions, and the multiple stakeholders involved.

This creates not only a level of richness through debate and discussion but also difficulty in pinpointing or discovering what exactly the social studies are. The social studies have also been problematic due to the tendency toward Eurocentric approaches. There remains much to be done about the limited perspectives still being presented,

It is often the case that the experiences of non-European Americans are treated especially poorly, with the coverage of the African American experience being notoriously bad. Research has shown that most high school U.S. history teachers, textbooks, and state and local curriculum standards routinely cover African American history in an incomplete and racially biased manner. (Thornhill, 2014, p. 2)

The social studies seem like an excellent point to examine and seriously consider the deleterious effects of that racism within our society but, in fact, this is not happening to the extent that would be truly transformative, and the dominant discourse remains domineering. Nelson and Pang (2014) focus on the nuances and possibilities which the social studies could contribute positively in problematizing race and racism:

> The social studies curriculum is the primary location in schools for inquiry into contemporary issues of prejudice. No other school subject has that civic mission... Basic principles and purposes of civic education and citizen development are stunted and distorted when discrimination against minorities remains a social norm. (pp. 203–205)

In addition, teachers have a role to play as argued by Fitchett, Starker, and Salyers (2012): "In social studies, cultural bias among practitioners is further revealed as teachers' unwittingly collude to propagate Eurocentric ideals due to a perceived fear that deviation from the prescribed curriculum will take away from instructional time" (pp. 587–588). This Eurocentrism is problematic for educators as well as the children they impact, in the ways in which it reifies whiteness and silences the damages done to those constructed as "others." The implications of these forms of racism for both white children and especially for children of color are destructive. Racism and white ethnocentricity continue to dominate the discourse and many of the educational practices, with resistance emerging even from preservice teachers in discussing challenging racial topics such as power, privilege, and whiteness (Garrett & Segall, 2013). Not only does this subtle poison persist in recreating these inequities and injustices but it also does so often insidiously.

To begin to challenge this, we can utilize the Critical Race Theory (CRT) assumption that racism is entrenched within American society (Delgado & Stefancic, 2012), therefore, in the educational systems and practices as well. Working with this assumption, we can begin to uncover some of the practices and possibilities for unveiling the silences and changing the approaches, practices, and inequitable education of our children. Without this, our children will continue to be miseducated, and this form of schooling will sustain the inequities and oppressions as well as the invisible privileges of the few. How can we hope to talk about "democratic citizens" if we can't discuss issues such as race and racism? This is especially vital given the current atmosphere involving repeated police violence against men, women, and children of color. In 2015, how do we reconcile occurrences, photos, and horrific stories reminiscent of the 1860s? The social studies offer a potential entry point for transformation.

HOPEFUL/CRITICAL PRACTICES

There is a danger in offering only critiques, with no potential for change, hence the necessity for hope. It is easy to condemn the wrongs, but potentially dangerous to do so with no alternatives present, as this leads to despair. Hope and despair are a delicate balance in an educational arena fraught with hypocrisies, damage, and neoliberalism.

In contrast, Freire (1998) connects hope with the naming of oppressions, and the subsequent denunciation and annunciation made possible through this:

> In this sense, the pedagogy that we defend ... is itself a utopian pedagogy. By this very fact it is full of hope, for to be utopian is not to be merely idealistic or impractical but rather to engage in denunciation and annunciation. Our pedagogy cannot do without a vision of man and of the world. (p. 492)

Creating and recreating our visions of pedagogy and schooling are inherently hopeful endeavors. Another aspect within the concept of hopeful/critical practices also involves elements of democratic power-sharing arrangements between students and teachers. I see the mutuality of agency as important here, with both students and teachers contributing to active resistance to racism within the social studies. Finding literature that engages the multiplicities which I am discussing (hopeful/critical, student/teacher agency, resistance to racism, and social studies) can be challenging, though there are certainly more articles than a decade ago. In order to begin analysis of this field and the concepts of resistance to racism with the social studies, multicultural education is a site which has contributed greatly to perspectives and understandings, and it is explored further below.

MULTICULTURAL EDUCATION

But multicultural education must go beyond the "contributions approach,"... Instead, I seek to move toward a "transformative" or "social action" approach to multicultural education. In transformative multicultural education [TME], the mainstream-centric perspective becomes only one of several presented. Various perspectives, frames, and content infuse students' understandings of the "nature, development, and complexity of U.S. society" (Banks, cited in Haviland, 2008, p. 41).

Nieto (2010) argues that multicultural education "is about *transformation*. I do not refer to just individual awareness but to a deep transformation on a number of levels—*individual, collective,* and *institutional*" (p.26). Multicultural education can go beyond additives to engage genuine personal and societal shifts.

Ironically, and in direct contrast to the continuing oppression that children are subjected to, movements for equity in the curricula (and in social studies) are not new (Bolgatz, 2007). With the multicultural education movement offering perspectives on reframing the discourse since the 1960s and 1970s (Takaki, 2008), and with CRT continuing to challenge the dominant discourses in multiple spaces (see Delgado & Stefancic, 2012; Gillborn, 2015; Ledesma & Caldéron, 2015), and concepts and theoretical frameworks such as Funds of Knowledge (González & Moll, 2002; González et al., 1995), it is difficult to reconcile the persistence of racism within our society and schools, except to acknowledge that racism is deeply entrenched within our society, and move forward after acknowledgement into activism against it.

Multicultural education can be used as a wider lens to begin to focus in on resistance to racism within the social studies. It deals broadly with issues of inequity (including most of the "isms") within education and advocates an attempt to reform not only the curriculum but also the institutions and the individuals operating within them. This is intended by surfacing the institutional factors such as power and privilege that continue to define the discourse and practices inequitably (Bolgatz, 2007). By shifting the institutions, the hope is that the ideal of education that is more equitable for all students involved will be achieved, regardless of ethnicity, race, gender, sexuality, or ability (Banks, 2004). There is a variety of standpoints and approaches within the field, which has been evolving over the past 30 years extensively, to the point which it is now considered a well-established field (Asher, 2007).

Scholars within this field come from a wide variety of perspectives and experiences. Some focus on the educational achievement of students of color and the redesigning of practices and curriculum through culturally relevant pedagogy (Gay, 2010; Howard, 2003a), while other authors point to multiple levels of transformation that would need to take place for equity to be achieved. This vision would involve work at all levels of education, and not simply changing the curriculum. This would suggest an intensive shift in the dominant discourse, attacking inequity from the many levels that it currently exists upon, rather than simply adding in curricular tokens that fail to examine inequity at deeper levels.

Darder (2012) engages curriculum, pedagogy, identity, and practice in her descriptions of "critical bicultural praxis." This approach involves taking multicultural education deeper and incorporating concepts such as hegemony and counter hegemony, voice, dialectical learning, and critical pedagogies including gender, location, and language. She offers examples of educators who harness critical bicultural pedagogies and shares their voices and insights into teaching more effectively and powerfully for students. This framework respects and engages not only the complexities of identity but also pedagogy and transformation.

Some see multicultural education as merely a superficial approach without the necessary critical components. As stated by Bolgatz (2007), "Unlike this basic version of multiculturalism, critical multiculturalism is concerned with broader issues of social power, economic systems, and the ways that difference is played out politically" (p.40), as echoed earlier by Nieto (2010). Bolgatz's definition points to one of the challenges with multicultural education and its power to address racism in a meaningful way, depth and challenge are necessary. Some scholars assert that multicultural education lacks the necessary criticality to create genuine change. Yosso (2002) suggests that multicultural approaches that simply focus on the "other," and offer cultures as unidimensional, which falls extremely short of genuinely addressing the problems. She advises readers that Critical Race Curriculum (CRC) offers an alternate possibility for action, where multicultural approaches can sometimes lag behind. Haviland (2008) shares this critique of multicultural education and stresses the importance of a social action component within the field. She argues: "Social-action multicultural education (SAME) builds on this foundation and goes further to equip and ignite students for social action and change" (p.41). Without the critical, change-oriented component, there can be the danger of oversimplifying difference and diversity and avoiding the deeper structural and power issues that maintain these gross social imbalances, hence replicating the status quo.

On the other hand, Ladson-Billings (2004) argues that in fact the field of multicultural education is dynamic and shifting and that the previously held definitions no longer apply:

Again, like jazz, multicultural education is less a thing than a process. It is organic and dynamic, and although it has a history rooted in our traditional notions of curriculum and schooling its aims and purposes transcend all conventional perceptions of education. (Banks & McGee Banks, 2004, p. 51)

As noted above, the field of multicultural education is immersed in debates about definitions and approaches to implementation. In some ways, it shares some of the difficulties and challenges of the social studies field; both are fields that are relatively new (multicultural education is newer), constantly in flux, and slightly difficult to define and implement, with plenty of debate. In other ways, these two fields are opposites—with social studies positioning (or claiming to position) itself to "create good citizens" and multicultural education responding by questioning: "Which citizens?" Whose citizenship and knowledge are taught and valued? And do these citizens share the same rights and enjoy the same privileges, or is there a clearly biased and imbalanced playing field? These types of questions need to be examined critically if we are trying to understand the ways inequity continues to be sustained within the social studies field. One of the ways we can start to delve more deeply into these types of questions is by examining different forms of resistance to racism within pedagogical and curricular approaches.

PEDAGOGICAL/CURRICULAR RESISTANCE

For some of us, just beginning to feel that our own stories are worth telling, the reminders of the "long dumb voices," the talk of "the rights of them the others are down upon" cannot but draw attention to the absences and silences that are as much a part of our history as the articulate voices, the shimmering faces, the images of emergence and success (Greene, 1993, p. 14).

This section covers a wide range of potential topics, and I chose to center my research on the readings that specifically touched on social studies practice in both K-12 settings as well as in higher education. I delved further into the literature and found (as in the quote above) that there were both presences and silences. I discovered that in fact my topics of interest had been partially covered by many people, but in different ways. As a result, my challenge was a bit like trying to become an expert weaver, or puzzle-constructor, to pull these disparate opinions and approaches together to form a cohesive whole. As mentioned above, it is vital that critical standpoints integrate reflections on and discussions of privilege, power, and status quo arrangements. There is a danger in some literature of implementing superficial "heroes and holidays" standpoints on multicultural education. This is true also of culturally relevant pedagogies, where despite good intentions, curricular practices that simplify and distort can become normative (focusing only once a year on the contributions of Black Americans, or celebrating two or three famous Latino(as)). These fail to engage with true change and are often unfortunately common.

In contrast, there is another approach which murmurs of transformation and challenges not only the content but also the process, the structure, and the ways knowledge is construed, as well as centering the experiences of those who have been forcibly placed on the receiving end of oppression. These were readings that touched on the ways that curriculum operates within a certain Eurocentric framework and larger institutional factors that help to replicate the status quo. Authors such as Ladson-Billings (2003), Ledesma and Caldéron (2015), Kumashiro, (2001), Yosso (2002), Cammarota (2007), Ruiz and Cantú (2013), Brown and Brown (2011), and Darder (2012) are concerned with disrupting the hegemonic practices through deeper examination of the ways that oppression occurs. These readings contained the criticality, but not always the hopeful aspects. I also noticed that many of them were from a teacher-directed position.

Self-reflection, self-development, engagement with community, and the humility to learn are all elements that need to be centralized for teachers as well as students when seeking transformative approaches. These are necessary in order to question the deep-seated and problematic and racist focus and foundations that so much of the curriculum operates from. Perhaps the discourse of equality and multicultural and anti-racist education has surfaced in the mainstream, but in such a way that the dominant discourse is not really threatened with genuine transformation. Perhaps "culture" (and work against inequity) has been colonized. Giroux (1997) states:

By analyzing culture uncritically either as an object of veneration or as a set of practices that embody the traditions and values of diverse groups, this view depoliticizes culture. More specifically there is no attempt to understand culture as shared and lived principles character- istic of different groups and classes as these emerge within inequitable relations of power and fields of struggle. (p.129)

This depoliticized version of multicultural education may be serv- ing the purpose of a band-aid solution to a much deeper wound. Below, I explore further critical versions of pedagogy and curriculum which may offer hope in our struggles toward equity in the classroom.

CRITICAL/TRANSFORMATIVE APPROACHES: IMPLICATIONS FOR K-12 AND HIGHER EDUCATION

"knowledge emerges only through invention and re-invention, through the restless, impatient, continuing, hopeful inquiry human beings pursue in the world, with the world and with each other." (Freire, 2000, p. 72)

Examples of the persistence of racism (as well as efforts to combat it in educational contexts) have expanded within the past decade. Ruiz and Cantú (2013) discuss the attacks on ethnic studies in Arizona, which involved the banning of books, the restriction of the curriculum focused on group identities and experiences, and the regulation of classrooms and teachers. The necessity of resistance is urgent, and their suggestions involve engaging preservice teachers in pedagogies of equity,

Both authors practice culturally relevant pedagogies and exemplify, through theirteaching, ways to use cultural frames and to practice decolo- nizing strategies when dealing with a diverse student body. (p. 78)

Ledesma and Caldéron (2015) thoroughly examine critical standpoints on education through the lens of CRT in both K-12 and higher educa- tion settings. Within the past decade, there has been a significant level of growth in the research and practices associated with these standpoints. These scholars present a review of the literature which incorporates effective practices in K-12 pedagogy (integration of meaningful counter- storytelling, Hip Hop pedagogies, examinations of interest convergence and historical revisionism). They also speak of the ways in which school policies and teacher biases and attitudes impact students negatively, in both K-12 and higher educational arenas.

Ladson-Billings (2003) edited a collection of essays that approach the social studies through a critical lens. A wide variety of authors cover a range of topics regarding the social studies while centering this discussion on the importance of race and racism. Howard (2003b) addresses this silence in the social studies by pointing to the fact that despite the claims of helping to "create good citizens," race and racism are persistently absent in the premier research journal, *Theory and Research in Social Education* (TRSE).

Silence carries meaning, and critical methodologies and pedagogies can work to shatter the silence and offer spaces for genuine thinking, genuine forms of democracy, discomfort, and potential growth. Ross, Mathison, and Vinson (2014) state:

> Social studies learning should not be about passively absorbing some-one else's conception of the world, but rather an exercise in creating a personally meaningful understanding of the way the world is and how one might act to transform that world. (p. 42)

Each of these pieces within this work point to the differing ways research-ers are challenging the silence about race and racism within the social stud-ies, and these offer hope in challenging the absences and centering active and critical practices to resist persistent curricular and societal racism.

Kumashiro mentions the fact that students come to school with only "par-tial knowledges" (p. 4), and then the school curriculum does not adequately surface or address these gaps in knowledge, and in fact may serve to reinforce them. In the social studies arena, he points to the focus of social studies:

> For example, when U.S. history curricula focus on political leaders, military conflicts, and industrial inventors, they are including the voices, experiences, and perspectives of only certain groups in society, namely, the privileged. Left silenced or pushed to the margins are such topics as immigration, the gendered division of labor and civil rights movements that can reveal the roles that the Othered in society have played in U.S. history. (p. 4)

He also offers critiques of the concept of the "add-in Other" by chal-lenging the fact that despite adding in other perspectives, we do not openly address or challenge the Norm by doing this. Kumashiro argues that in addition to this, because identities are multiple and interwoven complexities, the add-in approach can easily fall into oversimplification and essentialization. He states:

> It is easy to add difference to the curriculum in a way that complies with hegemony … we often expect that more voices added to the same story get the curriculum closer to telling a "truth" about World War II because we assume that truth is learned when all perspectives are told. (p. 6)

He continues to challenge this perception by reminding the reader of an extremely important postmodern tenet that all knowledge is closely tied to one's position and that, therefore, all knowledge is incomplete and only partially true. His advocacy of anti-oppressive education involves not only critical readings but also troubling the way that knowledge is presented to have meaning. This piece is also focused more broadly on issues of inequity, rather than solely on race, but still comes from a powerfully transformative approach.

Cammarota (2007) offers another perspective to social studies teaching and engaging in critical approaches and resistance to oppression through socially relevant curriculum. His focus in this piece is specifically on Latina/o students and he argues for critical consciousness to be developed through "active participation of Latina/o students in their lives, communities and futures" (p.87). The course that Cammarota developed incorporated itself into a preexisting social studies curriculum and focused on theoretical aspects of addressing racism such as CRT, critical pedagogy, and in addition involved practical, activist components; "they learned research methods including participant observation, interview techniques, photo documentation, and videography, for addressing the everyday injustices limiting their own and their peers' potential" (p. 89).

Cammarota's (2007) experience with implementing this challenging and also relevant curriculum proved successful. The students finished the educational goals, engaging in examinations of "cultural assimilation, critical thinking vs. passivity in education, racial and gender stereotypes of students and media representations of students of color" (p. 90). They also generated and analyzed data and created recommendations for the school board, superintendent, principal, teachers, and other community members. Cammarota argues that the findings of this study showed that those students who were involved felt that they were thinking differently about their own education and potential, as well as being helped to graduate from high school and consider the possibility of attending college. Cammarota states:

> Many activities in life, including education, become difficult undertakings for students constrained by severe social injustices. When these social injustices are engaged and critiqued, students begin to clear space for education. They become further engaged in learning when their education becomes a means by which they may challenge oppressive forces within their social contexts. (p.95)

In this situation, not only by surfacing the oppression of race within the curriculum but also by making it a focal point and opportunity for learning and self-empowerment, the youth involved in this work develop an ability to confront and potentially change those injustices woven into the tapestry of their daily lives.

One book that focuses on specific teaching strategies that relate broadly to social justice (and touches on racism within this framework) within the field of social studies was *Social Studies for Social Justice*. Wade (2007) seeks to approach social justice K-12 education by defining it through several different levels; she stresses care and fairness as focal points and elaborates on the other requirements for social justice education as student-centered, collaborative, intellectual, critical, multicultural, and activist. Though not solely dealing with race, this framework embraces several issues of injustice within the social studies classroom and the larger world. She stresses the importance of building a community in the classroom, and then of "reinventing" the curriculum. This form of micro resistance encourages taking and adapting the curriculum to follow social-justice-related themes. One of the most interesting facets of this book was the inclusion at the end of the work of a whole chapter on activism. This allows the social studies teacher and students to not only expand the classroom possibilities for justice but also reach outside of the classroom and to "become immersed in real-world problem solving and applying skills and knowledge learned in the classroom to civic action" (p.79).

Two more scholars who are important to examine when considering the implementation of critical standpoints are Howard Zinn and James

Loewen. Each has an extensive body of work which could be integrated into the classroom and/or serve as a space for teachers to engage with the social studies through multiple critical lenses. Specific works to be engaged could be Zinn's *A People's History of the United States* (2005), and *Howard Zinn on Race* (2011), as well as Zinn and Arnove's (2014) *Voices of a People's History of the United States*. There are a number of other works which are powerful and insightful as well, but these are examples of writings toward creating a more just and equitable society via reflections on our shared history and the power relations and implications involved in it. Loewen's work (2009, 2007) draws out the details of history through more multifaceted experiences and voices, and specific books that should be engaged include *Lies My Teacher Told Me* and *Teaching What Really Happened*. These book offer perspectives that are vital (and often ignored) based in the many voices of our country.

When examining resistance to racism within the larger context of the curriculum, certain theories can contribute to our understandings and activism. Though not in the social studies, Yosso (2002) uses CRT to examine not only racism but also the intersections with other forms of oppression through the curriculum. Her definition of curriculum is broad and insightful and includes "curricular structures, processes and discourses" (p.93). By using this multiple-leveled approach, she points to the interconnections between which types of classes present which kinds of knowledge, who is placed within those classes, and the discourses used to support these types of inequity. She uses CRT as a point of departure for a "Critical Race Curriculum" and argues that within this framework for five tenets to challenge racism through the curriculum:

(1) acknowledge the central and intersecting roles of racism, sexism, classism and other forms of subordination ... (2) challenge dominant social and cultural assumptions regarding culture and intelligence, language and capability, objectivity and meritocracy; (3) direct the formal curriculum toward goals of social justice and the hidden curriculum toward Freirean goals of critical consciousness; (4) develop counter-discourses through storytelling, narratives, chronicles, family histories, scenarios, biographies and parables that draw on the lived experiences students of color bring to the classroom; and (5) utilize interdisciplinary methods of historical and contemporary analysis to articulate the linkages between educational and societal inequality. (p.98)

This is a holistic and in-depth approach to examining profound ways that racism shapes the curriculum and potentially offers much for both scholars and teachers wishing to resist racism within their classrooms. It acknowledges the racist nature that educational institutions have been founded on and seeks to get to the deeper roots of the issues, rather than just touching them lightly and walking away.

In previous work (Daniels, 2011), I advocated for concrete strategies to incorporate critical perspectives and activities into the social studies. This involves utilizing resources such as CRT, Latino Critical Theory, and/or Tribal Critical Theory to enhance and deepen one's understanding as an educator first and then to engage more transformatively and deeply with the curriculum. For example, integrating discussions of property rights, poverty, colonization, linguistic discrimination through legal frameworks, as well as media outlets are all possibilities for integration. One possibility would be examinations of the practices of "red lining," where banks created zones within cities that prohibited the lending of money to certain people in certain areas (people of color). Though this is a historical situation, in the current atmosphere where police violence is predominant, certain discussions and critical examinations are necessary. We, as a society, are far from "done" with racism as evidenced by recent events in Texas, Baltimore, Ohio, New York City and Orlando. Racism continues to destroy individuals, lives, and our society.

CONCLUSION

If we are interested in truly transformative approaches, then we need to consider and grapple with the complexity of these issues in order to begin to ameliorate them, instead of supplying superficial acknowledgments. As stated by Brown and Brown (2011):

> social studies curriculum has served as a source of public controversy, concern, and debate within African American political and educational discourse. Second, of all the areas in social studies, U.S. history is an ideal space to substantively explore issues of race and social justice. (p. 9)

There are pieces which address racism and resistance within the social studies, but there are not as many as there could be, to provide useful resources and critiques of the dominant culture, engaging resistance powerfully with "road maps" for potential teachers to follow (even though perhaps maps themselves may be a challenge due to the uniqueness of contexts). As argued by Freire (2005) "education is a political act. Its non-neutrality demands from educators that they take it on as a political act" (p. 112). This call to action points to the ways in which our vocation of teaching is not a neutral or apolitical endeavor; we must choose where we stand and how we educate based on freedom or oppression. We need to engage with resistance.

To seek transformative resistance, we need to begin by addressing not only the curriculum but also the pedagogies, the teachers, and the structural aspects of racism. In seeking to change hegemony, to create citizens with power, critical perspectives, transformation is necessary; "reading the word is not only preceded by reading the world, but also by a certain form of writing it or rewriting it" (Freire, 1985, p. 18). The presence of literature which addresses resistance to racism in the social studies is a move in the right direction, but what about the absences and the silences? What do these say in terms of "creating good citizens"? As mentioned in the Pinar (1993) quote which opened this piece, we become "fragmented" if all that we learn is framed by distortion. To reduce this fragmentation, to move toward wholeness, we need to seriously address the structural, personal, and social limitations that we continue that recreate oppression, and we need to transform them. Perhaps the social studies are a site of transformation and resistance to racism, perhaps they can offer us this possibility if we choose to delve deeply into these challenges and honestly grapple with the issues. By grappling, perhaps our awareness and our reflection deepen, and perhaps then we emerge as activists, looking to support, engage, and challenge our students, our society, and ourselves.

KEYWORDS

- **Critical Multiculturalism**
- **Critical Race Theory**
- **Racism**
- **Resistance to Racism**
- **Social Studies Curriculum**

REFERENCES

1. Asher, N. (2007). Made in the multicultural USA: Unpacking tensions of race, culture, gender, and sexuality in education. *Educational Researcher, 36*(2), 65–73.
2. Banks, J., & McGee Banks, C.A. (Eds.). (2004). *Handbook of research on multicultural education* (2nd ed.). San Francisco, CA: Jossey Bass.
3. Bolgatz, J. (2007). More than Rosa Parks: Critical multicultural social studies in a fourth grade class. *Transformations, 18*(1), 39–51.
4. Brown, K.D., & Brown, A.L. (2011). Teaching K-8 students about race: African Americans, racism and the struggle for social justice in the U.S. *Multicultural Education, 19*(1), 9–13.
5. Cammarota, J. (2007). A social justice approach to achievement: Guiding Latina/o Students toward educational attainment with a challenging, socially relevant curriculum. *Equity & Excellence in Education, 40,* 87–96.
6. Daniels, E.A. (2011). Racial silences: Exploring and incorporating critical frameworks in the social studies. *The Social Studies, 102,* 211–220. doi: 10.1080/0037799 6.2011.558938.
7. Darder, A. (2012). *Culture and power in the classroom: Educational foundations for the schooling of bicultural students.* Boulder, CO: Paradigm Publishers.
8. Delgado, R., & Stefancic, J. (Eds.). (2012). *Critical Race Theory: An introduction* (2nd ed.). New York: NYU Press. Fitchett, P.G., Starker, T.V., & Salyers, B. (2012). Examining culturally responsive teaching in self-efficacy in a pre-service social studies education course. *Urban Education, 47*(3), 585–611.
9. Freire, P. (1985). Reading the world and reading the word: An interview with Paulo Freire. *Language Arts, 62*(1), 15–21.
10. Freire, P. (1998). The adult literacy process as cultural action for freedom. *Harvard Educational Review, 68*(4), 480–498.
11. Freire, P. (2000). *Pedagogy of the oppressed.* New York, NY: Continuum.
12. Freire, P. (2005). *Teachers as cultural workers: Letters to those who dare teach.* Boulder, CO: Westview Press.
13. Garrett, H.J., & Segall, A. (2013). (Re)considerations of ignorance and resistance in teacher education. *Journal of Teacher Education, 64*(4), 294–304.
14. Gay, G. (2010). *Culturally responsive teaching: Theory, research and practice.* (2nd ed.). New York, NY: Teachers College Press.
15. Gillborn, D. (2015). Intersectionality, Critical Race Theory, and the primacy of racism: Race, class, gender and disability in education. *Qualitative Inquiry, 21*(3), 277–287
16. Giroux, H. (1997). *Pedagogy and the politics of hope: Theory, culture and schooling.* Boulder, CO: Westview Press.
17. González, N., & Moll, L. (2002). Cruzando el Puente: Building bridges to funds of knowledge. *Educational Policy, 16*(4), 623–641. González, N., Moll, L., Tenery, M.F., Rivera, A., Rendon, P., Gonzales, R., et al. (1995). Funds of knowledge for teaching in Latino households. *Urban Education, 29*(4), 444–471.
18. Greene, M. (1993). The passions of pluralism: Multiculturalism and the expanding community. *Educational Researcher, 22*(1), 13–18.

19. Haviland, V.S. (2008). "Things get glossed over": Rearticulating the silencing power of whiteness in education. *Journal of Teacher Education, 59*(40), 40–54.
20. Howard, T. (2003a). Culturally relevant pedagogy: Ingredients for critical teacher reflection. *Theory into Practice, 42*(3), 195–202.
21. Howard, T. (2003b) The dis(g)race of the social studies: The need for racial dialogue in the social studies. In G. Ladson-Billings (Ed.), *Critical race theory perspectives on social studies: The profession, policies and curriculum* (pp. 27–43). Connecticut: Information Age Publishers.
22. Jorgensen, C.G. (2014). Social studies curriculum migration: Confronting challenges in the 21st Century. In E.W. Ross (Ed.), *The social studies curriculum: Purposes, problems and possibilities* (4th ed., pp. 3–23). Albany, NY: SUNY Press.
23. Kumashiro, K. (2001). "Posts" perspectives on anti-oppressive education in social Studies, English, mathematics and science classrooms. *Educational Researcher, 30*(3), 3–12.
24. Ladson-Billings, G. (Ed.). (2003). *Critical race theory perspectives on social studies: The profession, policies and curriculum.* Connecticut: Information Age Publishers.
25. Ladson-Billings, G. (2004). New directions in multicultural education. In J. Banks & C.A.McGee Banks (Eds.), *Handbook of research on multicultural education* (pp. 50–65). San Francisco: Jossey Bass.
26. Ledesma, M.C., & Caldéron, D. (2015). Critical Race Theory in education: A review of past literature and a look to the future. *Qualitative Inquiry, 21*(3), 206–222. doi: 10.1177/1077800414557825.
27. Lee, J.K. (2005). Reconsidering the debate: Social studies, history and academic disciplines. *International Journal of Social Education, 20*(1), 60–63.
28. Loewen, J. (2007). *Lies my teacher told me: Everything your American history textbook got wrong.* New York, NY: Touchstone.
29. Loewen, J. (2009). *Teaching what really happened: How to avoid the tyranny of textbooks and get students excited about doing history.* New York: Teachers College Press.
30. Nelson, J.L., & Pang, V.O. (2014). Prejudice, racism and the Social Studies curriculum. In E.W. Ross (Ed.), *The social studies curriculum: Purposes, problems and possibilities* (4th ed., pp. 203–225). Albany, NY: SUNY Press.
31. Nieto, S. (2010). *The light in their eyes: Creating multicultural learning communities* (10th anniversary edition). New York: Teachers College Press.
32. Pinar, W. (1993). Notes on understanding curriculum as racial text. In C. McCarthy & W. Crichlow (Eds.), *Race, Identity & Representation in Education* (pp. 60–70). New York: Routledge.
33. Ross, E.W., Mathison, S., & Vinson, K. (2014). Social Studies curriculum and teaching in an era of standardization. In E.W. Ross (Ed.), *The social studies curriculum: Purposes, problems and possibilities* (4th ed., pp. 25–49). Albany, NY: SUNY Press
34. Ruiz, E.C., & Cantú, N.E. (2013). Teaching the teachers: Dismantling racism and teaching for social change. *Urban Review, 45*, 74–88. doi: 10.1007/s11256-012-0225-2.

35. Takaki, R. (2008). *A different mirror: A history of multicultural America* (Revised edition). New York: Brown and Company.
36. Thornhill, T.E. (2014). Resistance and assent: How racial socialization shapes Black students' experience learning African American history in high school. *Urban Education*, 1–26.
37. Wade, R. (2007). *Social studies for social justice: Teaching strategies for the elementary classroom.* New York: Teachers College Press.
38. Yosso, T. (2002). Toward a critical race curriculum. *Equity & Excellence in Education, 35*(2), 93–107.
39. Zinn, H. (2005). *A people's history of the United States.* New York: Harper Perennial.
40. Zinn, H. (2011). *Howard Zinn on race.* New York: Seven Stories Press.
41. Zinn, H., & Arnove, A. (2014). *Voices of a people's history of the United States* (10th Anniversary edition). New York: Seven Stories Press.

CHAPTER 9

"BUT HOW DO I KNOW IF IT IS AUTHENTIC?" EXAMINING PICTURE BOOKS REPRESENTING CHINESE CULTURE

YONG YU*

State University of New York at Plattsburgh, 101 Broad St, Plattsburgh, NY 12901, United States

ABSTRACT

Multicultural children's literature has the power to perpetuate or disrupt biases and stereotypes, depending on how authentically it reflects the culture depicted. Many teachers struggle with determining whether a multicultural children's book is culturally authentic or whether it is appropriate to use in the classroom. As a literacy professor born and raised in China, the author examined 87 fiction picture books about Chinese/Chinese-American culture in this study, hoping to offer an insider view about the books for classroom teachers. Analysis found that nearly 50% of the books reviewed in this study reflect Chinese/Chinese-American culture highly authentically. However, inaccurate cultural facts and stereotypes are still present in 20% of the books, including several recently published titles. The author suggests that if the purpose of using multicultural children's

*Author note: I would like to thank Dr. Anne Creany, professor emerita of literacy education at Indiana University of Pennsylvania, who sparked my interest in children's literature and shared her insight during the course of this research. I would also like to thank my colleagues whose suggestions and comments have greatly improved the manuscript.

literature is to develop healthy cultural identity and dispel stereotypes, it is important for teachers to not only expose students to books with a high level of cultural authenticity but also to teach them what to look for and why a book is not authentic so as to foster critical literacy skills and multicultural awareness.

INTRODUCTION

A new student has arrived in Ms. Kingsley's second-grade classroom, and she wants to make him feel welcome. Since the student is Chinese American, Ms. Kingsley checks her classroom library for books that reflect Chinese/Chinese American culture. She finds Tikki Tikki Tembo, *a book beloved by her students for its tongue-twisting refrain. However, Ms. Kingsley feels uneasy about the basic premise behind the supposed folktale—that Chinese parents now give their children short names because long ago Tikki Tikki Tembo No Sa Rembo Chari Beri Ruchi Pip Perri Pembo almost drowned because his name was so long that it was difficult to summon help. Ms. Kingsley prepares to search for other books in the school library. However, she wonders how she can tell whether a book represents Chinese/Chinese American culture authentically.*

Many teachers in the United States may find themselves in a position similar to Ms. Kingsley. During the past three decades, racial and ethnic diversity in the United States has grown dramatically. According to the Federal Interagency on Child and Family Statistics (Forum on Family and Child Statistics, 2014), minority children account for 47% of children under 18 in the United States. Of them, 5% are children of Asian descent. The Asian population increased by 43% between 2000 and 2010, faster than any other race group. Among Asian populations, Chinese is the largest detailed group, constituting approximately 23% of Asian-American individuals (U.S. Census Bureau, 2010).

This demographic change in our classrooms demands that "all children are taught about the other children growing up in this world" (Suh & Samuel, 2011, p. 3). On the one hand, children need to see themselves and people who look like them reflected in the books they read. If children never see themselves in books, they are learning that they are not important (Gangi, 2008). On the other hand, children need to read books about people whose cultures are different from their own. They need to develop the awareness that their worldview is not the only one, but one of many perspectives. Multicultural children's literature has the power to fulfill both purposes (Perini, 2002). It "helps children identify with their own culture, exposes children to other cultures, and opens the dialogues on issues regarding diversity" (Colby & Lynon, 2004, p. 24). Through reading and bonding with the characters in the stories, children learn to understand, respect, and empathize with children from diverse backgrounds (Klefstad & Martinez, 2013).

Although multicultural children's literature can be a great learning tool, it has the power to either continue or disrupt stereotypes, depending on whether it represents the culture authentically (Colby & Lyon, 2004). Culturally authentic literature can break down negative stereotypes and encourage understanding and appreciation of different cultures (Noll, 2003). If children are exposed to literature that is not authentic or true to the culture it depicts, the multicultural benefits can be diminished, be ineffective, or even be negatively affected (Smith & Wiese, 2006). In addition, children's identities and understandings are negatively influenced when "literature contains misinformation and warped images" (Noll, 2003, p. 182). Inauthentic Chinese folktales cause problems when passed down to Chinese-American children who need to understand their ancestral heritage, as well as American children in search of understanding Chinese/Chinese-American culture (Dong, 2013). Therefore, a teacher's ability to select children's literature with authentic cultural representation has significant impact on its power to bridge cultural gaps in the classroom.

Classroom teachers in the United States, however, are poorly prepared to evaluate cultural authenticity of multicultural children's books. Many teachers who feel confident selecting literature on the basis of curricular goals or literary criteria lose that confidence when it comes to selecting multicultural literature, either fearing being accused of insensitivity or feeling a lack of knowledge about cultural groups other than their own (Bishop, 1993). When asked to identify two children's books, the majority of the pre- and in-service teachers in Brinson's (2012) study were only able to name books depicting White American, and none about any minority groups. This lack of professional preparation might have contributed to

Loh-Hagan's (2014) findings about the lack of demand for high-quality Asian-American books from teachers, librarians, and parents, in spite of the availability. The purpose of this study is to examine the authenticity of children's picture books representing Chinese/Chinese-American culture and to provide a resource guide for pre- and in-service teachers to select multicultural Children's literature.

CULTURAL AUTHENTICITY

Whether a children's book is considered culturally authentic has much to do with how authenticity is defined. The most common understanding of cultural authenticity is probably the reader's sense of truth in how a specific cultural experience has been represented within a book, particularly when the reader is an insider to the culture portrayed in that book (Howard, 1991). Since the act of reading literature involves a transaction between the reader and the text, each is unique and results in different interpretations (Rosenblatt, 1995), and given the range of experiences within any cultural group, this definition of cultural authenticity sets the stage for debates about the authenticity of a particular book (Short & Fox, 2003). Some researchers suggest cultural authenticity refers to accurate and non-stereotyped representation of characters, settings, values, and beliefs of a culture (Harada, 1995; Louie, 2006; Nilsson, 2005; Wilfong, 2007). Others argue that cultural authenticity is not just accuracy or the avoidance of stereotypes but involves cultural values and issues/practices that are accepted as norms of the social group (Bishop, 2003; Mo & Shen, 2003; Yokota & Kolar, 2008).

At the center of these discussions is the insider/outsider debate. Some children's authors and researchers believe that a culture insider is more likely to give an authentic view of what members of that group believe to be true (Bishop, 2003; Yokota, 1993; Yokota & Kolar, 2008). Others see this distinction as a form of censorship and an attempt to restrict an author's freedom to write (Lasky, 2003). Cai (2003) argues that the question reflects larger issues of power structure—a power struggle over whose books get published. Bishop (2003) points out that the debate focusing on cultural membership is simplistic and overlooks the broader sociocultural issues.

Such issues are discussed by the chapter authors in the book edited by Short and Fox (2003) and summarized as below: (1) Can outsiders write authentically about another culture?(2) How does an author's social responsibility relate to authorial freedom?(3) How does cultural authenticity relate to literary excellence in evaluating a book? (4) What kinds of experience matter for authors in writing culturally authentic books?(5) What are an author's intentions for writing a particular book? (6) What are the criteria beyond accuracy for evaluating the cultural authenticity of the content and images of a book? And finally, (7) what constitutes an "insider" perspective on cultural authenticity?

The current study follows Bishop's (2003) two-dimensional framework that defines cultural authenticity as "the extent to which the work reflected the cultural perspective or worldview of the people whose lives are reflected in the work" (2003, p. 28), and the accuracy of authenticating details such as characters' language and cultural information possessed by members of a cultural group. In this study, the insider perspective is characterized by what Cai refers to as a "special sense of reality" .... "A Chinese sense rather than an American sense," (which is) "not inherited through genes but acquired through direct and indirect experiences" (2003, p. 172).Accuracy is used to refer to cultural facts (Mo & Shen, 2003)—the visible culture, the part of an iceberg shown above the surface. Authenticity deals with values and beliefs (Mo & Shen, 2003)—the invisible culture, the hidden section of an iceberg below the surface. Accuracy is an essential but not a sufficient condition for a multicultural children's book to be authentic.

Bishop (1993) suggests that one should consider the underlying purpose and diverse points of view of multicultural children's literature in order to evaluate its cultural authenticity. She placed children's fiction and picture books into three categories: specific, generic, and neutral. A culturally specific children's book illuminates the experience of a member of a particular, non-white cultural group, by detailing the specifics of daily living that will be recognizable to members of that group. A generic multicultural children's book features characters who are members of so-called minority groups, but contains few specific details that might serve to define those characters culturally. A culturally neutral book seems to feature people of color, but is fundamentally about something else. Bishop argues that the generic and neutral books can be evaluated "on the basis of their accuracy, their literary and visual artistry, and possible omissions, but cultural authenticity is not likely to be a major consideration" (p. 46).

STUDIES EXAMINING AUTHENTICITY OF BOOKS ABOUT CHINESE CULTURE

Several studies have examined cultural authenticity of children's books about Chinese/Chinese-American culture using relatively large samples (Cai, 1994; Lin, 1999; Liu, 1998; Liu, 1993). While the studies concluded that most books represented Chinese/Chinese-American culture authentically, each found issues regarding accuracy and stereotypes. Mei-Ying Liu's (1993) study of 34 Chinese fiction books for children published from 1925 to 1991 found unauthentic use of language and inaccurate information in 54% of the books. Cai (1994) examined 73 children's books portraying Chinese/Chinese-American culture and found that many visual images in these books are not culturally authentic, resulting in vulgarizing the integrity of Chinese culture or ambiguity of cultural identity. Li Liu (1998) studied 57 children and adolescent books about Chinese/Chinese-American culture published between 1980 and 1997. The findings indicate misrepresentation of Chinese people in such aspects as physical features, clothes, occupation, gender value, and lifestyle. Lin (1999) studied 30 children's books featuring Chinese-American female protagonists published from 1963 to 1997. The findings suggest that 22 out of the 30 books promoted gender stereotypes of traditional Confucian ideology, which expects women to exercise their love in an altruistic form due to the need for men to embody the male-dominated power and privilege. Two findings common to the four studies are that books published most recently are better than those published earlier; and books written and illustrated by Chinese/Chinese-American authors and artists are less flawed by cultural accuracy than those by non-Chinese authors/illustrators.

All of the aforementioned studies were conducted in the 1990s. More recent studies on children's literature about Chinese/Chinese-American culture either focused on a specific aspect or were conducted as part of research on literature of multiple cultures. For example, Hsieh and

Matoush (2012) compare and contrast the cultural authenticity in several versions of Mulan. Dong (2013) discussed cultural authenticity of folktales, particularly concerns about the extent to which a retold or imagined folktale is authentic to the social and moral values of the Chinese/Chinese-American community it is representing. Eun Young and colleagues' (2014) collaborative analysis of cultural authenticity in multicultural children's books included 10 books representing Chinese culture. Results of these studies reveal a consistent pattern that the majority of the books represent Chinese/Chinese-American culture authentically in both texts and illustrations, but stereotypes and inaccuracies are still present.

Review of the literature has found no study examining authenticity of children's books about Chinese/Chinese-American culture published since 2000 using a large sample. In addition to the lack of information about cultural authenticity of children's books published in recent years, books like *Tikki Tikki Tembo* and *Five Chinese Brothers*, which have been widely recognized as falsely representing Chinese culture, are still reprinted and circulating. Teachers still struggle with selecting children's books that represent authentic portrayals of Chinese/Chinese-American individuals and culture. The present study intends to bridge this gap by examining picture books available to children and teachers from an insider's perspective and providing a synthesized guideline for teachers to apply when selecting multicultural books for children.

THE STUDY

The study was conducted in a small city of about 20,000 residents in upstate New York. I surveyed three local elementary school libraries and the city public library. Books were selected according to the following criteria: (1) books written by and about Chinese or Chinese Americans. I adopted the criteria used by the Cooperative Children's Book Center (CCBC) at University of Wisconsin–Madison (2014) and considered a book as "about" if the main character/subject is Chinese or Chinese Americans; (2) picture book, "a cohesive text in which narrative and art work together to convey meaning and mood" (Fountas & Pinnell, 2006, p. 126); (3) fiction, "a narrative that is imagined rather than real" (Fountas & Pinnell, 2001, p. 393); and (4) main character is a human being instead of an animal.

The checklist used to evaluate the books in this study was created by synthesizing the criteria and checklists available in the related literature. It covers multiple categories, including character, setting, language, theme, cultural details, illustration, authorship and cultural perspectives, and citation and acknowledgment (See Table 9.1). The guiding questions have been influenced by suggestions in the works of Bishop (1993), California State Department of Education (1998), Hancock (2000), Harada (1995), Hearne (1993a, 1993b) ,Higgins (2002), IRA Notable Books for a Global Society Committee (n.d.), Louie (2006), Norton (2009), Wilfong (2007),and Yokota (1993). Although I used the checklist to evaluate cultural authenticity of books representing Chinese/Chinese-American culture, it is applicable to any specific culture with minor adaptations.

I evaluated every book in this study by examining each of these areas, asking and responding to the guiding questions shown in the table. In the course of analyzing the books, I drew upon my knowledge about and experiences of living in China from birth to age 40, and pursued degrees in higher education in both China and the United States. When authenticity was questionable, authorship and citations were researched to help make the decision.

PROFILE OF THE BOOKS EXAMINED IN THE STUDY

A total of 87 books met the inclusion criteria and were selected for this study. Among them, 38 were written by Chinese or Chinese Americans and 49 by non-Chinese authors. In addition, 45 books are traditional literature, including folktales and legends, 30 are realistic fictions, 10 are historical fictions, and 2 are fantasies. Figure 9.1 shows the percentage of genre.

Figure 9.2 shows the number of books published in each decade from 1930s to 2010s. Overall, traditional fiction in the sample was almost the only genre published before 1980s, which increased significantly through 1980s and 1990s, but decreased dramatically after 2000. In contrast, picture books of realistic fiction were almost nonexistent until the 1990s but increased quickly after 2000. Only four books in this study were published after 2010, including three realistic fictions and one historical fiction, all written and illustrated by Chinese/Chinese Americans.

TABLE 9.1 Checklist for Evaluating Cultural Authenticity

Category		Guiding Questions
Authenticity	Character	• Are the characters believable, fully developed, and portrayed in depth? • Are the names of characters culturally authentic? • Do people of color lead as well as follow? • Do they solve their own problems or depend on white benefactors? • Are the members of a cultural group included for a purpose rather than to fulfill a "quota"? • Are characters described accurately and free from stereotypes?
	Setting	• Is the setting natural in relation to the content? • Is the setting consistent with a historical or contemporary time, place, or situation of a particular culture? • Is the factual information describing a historical setting accurate in detail? • Are minorities depicted exclusively in ghettos and ethnic enclaves (e.g., Chinatown)?
	Language	• Is the language of the characters accurate, authentic, and appropriate to their historical time, educational and social backgrounds, and social situations in which they are operating? • Is the writing used to express the language of a culture made to look exotic? • Is dialect natural and blended with plot and characterization instead of perceived as a substandard or inferior form of language?
	Theme	• Is the theme consistent with the values, beliefs, customs, traditions, and conflicts of the specific cultural group? • Is the theme universal to all cultures and applied appropriately to the culture portrayed in a way that a cultural insider would do?
	Cultural details	• Do the details of the story help the readers gain a sense of the culture they are reading about? • Are the details a natural part of the story, giving nuances of daily life, rather than appearing as though the purpose of the book is to explain cultural details?

TABLE 9.1 Continued

Category	Guiding Questions
Illustration	• Do illustrations present accurate, natural, non-stereotypical images of people from the culture depicted? • Do they show variety in physical features among the individuals of any one group, or do they all look alike? • Do minority characters look like white people except for being tinted or colored?
Authorship and cultural perspectives	• Is the author an insider or outsider of the cultural group portrayed? • Whose perspectives and experiences are portrayed? • What is the range of insider perspective? • How do the author's experiences connect to the setting and characters in this book? • Can readers get a sense of the experiences and/or research on which the book is based?
Citations or acknowledgements	• Does the author cite or acknowledge works or people that contributed to the knowledge base used for the writing of the book? • Does author of a folktale cite the source and explain the adaptations?

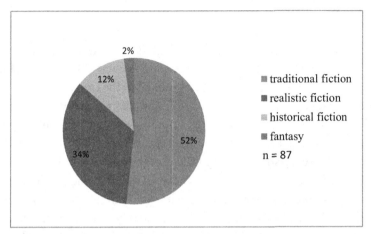

FIGURE 9.1. Percentage of children's picture books portraying Chinese/Chinese-American culture by genre. This figure depicts the percentage of each genre of picture books. They were found in three elementary school libraries and one public library.

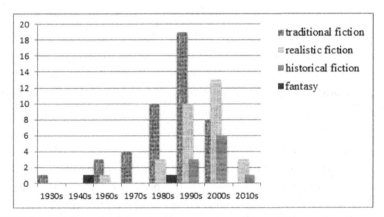

FIGURE 9.2 Children's picture books portraying Chinese/Chinese-American culture by year of publication. This figure illustrates picture books found in this study that were published in each decade since 1930s.

FINDINGS AND DISCUSSIONS

The findings of cultural authenticity of the 87 books are presented in this section by genres.

FOLKTALES

Forty-five books included in this study are folktales. Among them, 15 were written by Chinese/Chinese-American authors and 29 by non-Chinese authors. Table 9.2 lists the 45 folktales ranked in the order of high, moderate, and low level of authenticity according to the results of analysis.

Of the 45 books, 22 demonstrated a high level of authenticity regarding character, setting, language, theme, cultural details, illustration, and represented Chinese perspectives regardless of the author's cultural membership. The illustrations successfully portrayed positive and realistic images of Chinese people that represented a variety of ethnic groups and historical periods. Several illustrators used classical Chinese art styles including brush and ink (e.g., *The Magic Horse of Han Gan* [Chen, 2006]), paper cut (e.g., *The Story of Chopsticks* [Compestine & Xuan, 2001]), and ink and watercolor painting on handmade rice paper (e.g., *The Jade Stone* [Yacowitz & Chen, 1992]). Both content and illustration in these 29 stories are free from stereotypes. Nineteen books provided some sort of notes that explained the source and/or background of the stories. Four books, *Liang and the Magic Paintbrush* (Demi, 1980), *The Long-Haired Girl* (Rappaport &Yang, 1995), *The Boy Who Swallowed Snakes* (Yep, Tseng, & Tseng, 1994), and *The Emperor and the Kite* (Yolen & Young, 1967), did not cite any source, but did not raise series concerns about authenticity.

Twelve books of folktale showed moderate authenticity and demonstrated problems such as lack of depth treatment of culture and inaccurate cultural facts. This group included six titles by Demi. These stories reflected themes that were universal, for example, honesty and kindness, but did not offer rich cultural details that provided the reader a sense of Chinese culture except the characters' name and the illustration. The magic tapestry: a Chinese folktale was weak at character development. Nothing was mentioned about what happened to the two older sons at the end of the story. *The Empty Pot* (Demi, 1990) and its sequel, *The Greatest Power* (Demi, 2004), both had quality story lines and illustration. However, the fact that Ping, the boy who was not a member of the royal family, was chosen to be the emperor would not happen in China because of the system of hereditary succession. In fact, *The Empty Pot* is a Korean folktale currently included in the first-grade Chinese textbooks in China as an example of Korean literature.

TABLE 9.2 Cultural Authenticity in Picture Books of Folktales

Level of Authenticity	Title	Author	Illustrator	Year
High	The Lost Horse: A Chinese Folktale	Ed Young*		1998
	Night Visitors	Ed Young*		1995
	The Hunter: A Chinese Folktale	Mary Casanova	Ed Young*	2000
	The Magic Horse of Han Gan	Chen Jiang Hong* Translated by Claudia Xoe Bedrick		2006
	Ten Suns: A Chinese Legend	Eric Kimmel	YongSheng Xuan*	1998
	Eyes of the Dragon	Margaret Leaf	Ed Young*	1987
	Yeh-Shen : A Cinderella Story from China	Retold by Ai-Ling Louie*	Ed Young*	1982
	The Cricket Warrior	Retold by Margaret & Raymond Chang*	Warwick Hutton	1994
	The Beggar's Magic: A Chinese Tale	Retold by Margaret & Raymond Chang*	David Johnson	1997
	White Tiger, Blue Serpent	Grace Tseng*	Jean & Mou-sien Tseng*	1999
	The Jade Stone	Caryn Yacowitz	Ju-Hong Chen*	1992
	The Man Who Tricked a Ghost	Laurence Yep*	Isadore Seltzer	1993
	Lon Po Po: A Red-Riding Hood Story from China	Ed Young*		1989
	The Story of Chopsticks	Ying Chang Compestine*	YongSheng Xuan*	2001
	The Runaway Rice Cake	Ying Chang Compestine*	Tungwai Chau*	2001
	The Real Story of Stone Soup	Ying Chang Compestine*	Stephane Jorisch	2007
	The Dragon's Pearl	Julie Lawson	Paul Morin	1993

TABLE 9.2 Continued

Level of Authenticity	Title	Author	Illustrator	Year
	The Seven Chinese Brothers	Margaret Mahy	Jean & Mou-sien Tseng*	1990
	White Wave: A Chinese Tale	Diane Wolkstein	Ed Young*	1979
	The Voice of the Great Bell	Lafcadio Hearn	Ed Young*	1989
		Margaret Hodges		
	Liang and the Magic Paintbrush	Demi		1980
	The Long-Haired Girl: A Chinese Legend	Doreen Rappaport	Yang Ming-Yi*	1995
	The Boy Who Swallowed Snakes	Laurence Yep*	Jean & Mou-Sien Tseng*	1994
	The Emperor and the Kite	Jane Yolen		1967
Moderate	Chenpingand his Magic Axe	Demi		1987
	The Magic Boat	Demi		1990
	The Greatest Treasure	Demi		1998
	The Magic Tapestry: A Chinese Folktale	Demi		1994
	The Empty Pot	Demi		1990
	The Greatest Power	Demi		2004
	Legend of the Milky Way	Jeanne Lee		1982
	The Weaving of a Dream: A Chinese Folktale	Marilee Heyer		1986

TABLE 9.2 Continued

Level of Authenticity	Title	Author	Illustrator	Year
	Ma Lien and the Magic Brush	Retold by Hisako Kimishima; Translated by Alvin Tresselt	Kei Wakana	1968
	8,000 Stones: A Chinese Folktale	Diane Wolkstein	Ed Young*	1972
	Dragon Prince: A Chinese Beauty & Beast Tale	Laurence Yep*	Kam Mak*	1997
	Monkey King	Ed Young*		2001
Low	Once There Were No Pandas: A Chinese Legend	Margaret Greaves	Beverley Gooding	1985
	The Magic Wings: A Tale From China	Diane Wolkstein	Robert Andrew Parker	1983
	The Paper Dragon	Marguerite Davol	Robert Sabuda	1996
	Everyone Knows What a Dragon Looks Like	Jay Williams	Mercer Mayer	1976
	Min-Yo and the Moon Dragon	Elizabeth Hillman	John Wallner	1992
	The Willow Pattern Story	Allan Drummond		1992
	The Seeing Stick	Jane Yolen	Remy Charlip & Demetra Maraslis	1977
	Tikki Tikki Tembo	Arlene Mosel	Blair Lent	1968
	The Five Chinese Brothers	Claire Huchet Bishop	Kurt Wiese	1938

*Chinese or Chinese American.

Each of the remaining six books either showed minor flaws of authenticity or contained adaptation that might deviate too far away from Chinese culture. *Legend of the Milky Way* (Lee, 1982) changed the story line but did not explain the reason. The illustrations in *The Weaving of a Dream: A Chinese Folktale* (Heyer, 1986) demonstrated a noticeable intention to create an exotic impression. Kimishima's retelling of *Ma Lien and the Magic Brush* (1968) was translated from a story originally published in Japan. In this version, the mandarin's robe had the pattern of double happiness, a decoration only for wedding instead of being printed on daily clothes. In *8,000 Stones: A Chinese Folktale* (Wolkstein & Young, 1972), the author mistakenly named the main character Cao Pei (189–226 A.D.) instead of Cao Chong (196–208 A.D.), though both were sons of Cao Cao, the first emperor of Kingdom Wei (220–265 A.D.). *Dragon Prince* (Yep & Mak, 1997) was a story set at modern China as was indicated by the clothes of the main characters. However, the Dragon Prince was dressed as someone from the Qing Dynasty (1636–1912). *Monkey King* (Young, 2001) was a wonderful retelling of the beginning part of a Chinese classic epic. In the story, Ed Young painted the Buddhist monks' robe with colorful flower patterns, which was a complete deviation from the red and/or yellow patchwork in Buddhism culture in China. When analyzing these books, I examined the source or background notes and introduction about the authors and illustrators, hoping to determine whether these inaccuracies were due to lack of research. However, as most of these 12 books did not provide such information, teachers are recommended to be aware of these flaws when selecting the books.

The last nine books of folktales showed a low level of authenticity and had major flaws in multiple areas. Stereotypes and biases were present in some of them. Three dragon stories were categorized into this group. All three were original folktales, or literary folktales, which are stories that come from the author's imagination rather than the culture depicted. *The Paper Dragon* (Davol & Sabuda, 1996) depicted a dragon as evil and brutal, breathing out fiery fire, and living in a cave—a typical European perspective. In Chinese culture, dragons are symbols of royal power and protection, often associated with water. In addition, the protagonist in the story was portrayed as a person in the Qing Dynasty dressed in long gown and wearing a queue, while other characters' clothes and hairstyles were from earlier ages in Chinese history. *Min-Yo and the Moon Dragon* (Hillman & Wallner, 1992) was another literary tale that misrepresented Chinese dragon as an animal with blue eyes, golden wings, and a body of dinosaur. The story line was also misleading because the idea of the dragon living on the moon was unknown in China. Instead, the moon is the habitat of the beautiful moon fairy Chang-e. *Everyone Knows What A Dragon Looks Like* (Williams & Mayer, 1976) was also a literary tale related to dragon. It was set in Wu, an imaginary village in China. In spite of the depth of the theme, there was no way I can tell the story happened in China just by looking at the illustrations—the dragon was depicted as fierce, the houses looked like those in Western fairy tales, and the characters looked very alien with big noses and red cheeks.

Four other original folktales in this group also presented issues about authenticity in varying degrees. *The Willow Pattern Story* (Drummond, 1992) was a tale invented by people in Europe to explain the willow pattern plate which began to be popular in the 1800s in England. Even though the story was set in China, the characters and architecture looked more European than Chinese. *The Seeing Stick* (Yolen, Charlip, & Maraslis, 1977) contained inaccuracy in both content and illustration. The information about the naming tradition in China in the background note was inaccurate. The images of the characters did not reflect realistic Chinese people. The emperor had blue eyes and wore clothes that looked like Arabic.

Regardless of the inaccuracies and unauthenticity, none of these five tales are comparable to *The Five Chinese Brothers* (Bishop & Wiese, 1938) and *Tikki Tikki Tembo* (Mosel & Lent, 1968) in terms of their negative and racist influences. *The Five Chinese Brothers* contained stereotypical portrayals of the characters, an absurd and negative message, and a racist overtone within the printed text. All Chinese characters were depicted as exactly alike with bucked teeth, slanted and slit eyes, yellow skin, and long queues. *Tikki Tikki Tembo* reflected ignorance about Chinese culture and made racist claims about Chinese naming tradition. As a matter of fact, I was surprised that both books were still available in all of the three local elementary school libraries. The Asian-American community called for a ban on the book for its inauthenticity and racist Orientalism in the 1960s–1970s, but failed (Dong, 2013). Up to today, both books continue to be published and read as "classics." The back cover of the edition *Tikki Tikki Tembo* I examined stated there were "over 1,000,000 copies in print."

Two Chinese folktales in this group are retelling of Chinese legends. Both contained inaccurate facts and stereotypes, and neither provided the reader a specific sense of Chinese culture. According to the note on the back cover, *Once There Were No Pandas: A Chinese Legend* (Greaves & Gooding, 1985) was "a retelling of a Chinese legend" about the origin of the panda. Although there is a similar legend in the Tibetan area in China, the current retelling was not authentic and lacked deep treatment of culture. The only human character in the story had slanted eyes and the same facial expression throughout the story. The Chinese words, 观音 (Guānyi), a *bodhisattva*, was spelled as光阴 (guāngyi), which means *time*. *The Magic Wings: A Tale from China* (Wolkstein & Parker, 1983) had a fine print note, stating that the story was based in part on the story line of Growing Wings from *The Milky Way and Other Chinese Folk Tales* (Lin, 1961). While the authenticity of the original source needs further research, I was concerned about the book's authenticity when seeing the characters wore clothes of different historical times that were hundreds years apart. The image of Spirit in Heaven Who Grows Wings in the story resembled the fairies in Western fairy tales rather than Chinese folklores.

A noticeable pattern about the folktales in this study is that the majority of folktales reflecting high level of authenticity (92%) were either authored or illustrated by Chinese/Chinese Americans. (See Table 9.2). This is consistent with the existing literature that authors who are members of the culture depicted are more likely to represent that culture authentically (Cai, 2003). Another pattern is the absence of source notes. Only one-third of folktales being reviewed contained source notes that Hearne (1993a) would consider as "well-made" or "model". Other books had either "fine-print" source notes (11%), a note mentioning the book is a retelling of Chinese tale, or no source notes at all (53%). Moreover, more than half of the folktales (58%) that did not have acceptable source notes showed serious issues of authenticity in character, language, theme, perspective, and/or illustration. If a picture book of folktale has no information about the source and adaptation of the story, readers who are conversant with Chinese/Chinese-American culture may rely on their background knowledge to determine to what extent it reflects the culture authentically.

However, it is challenging for readers with insufficient background knowledge and/or little interaction with Chinese/Chinese-American culture to make an accurate decision.

REALISTIC FICTION

Thirty picture books included in this study are realistic fiction. Out of the 30 books, 16 are written by Chinese/Chinese Americans and 14 are written by authors who are not Chinese. Based on the results of the analysis, 13 books represented Chinese/Chinese-American culture authentically, 10 books showed moderate authenticity, and 7 showed poor authenticity. Table 9.3 presents these books in the order of high to low authenticity.

All 13 books that are considered highly authentic reflect specific Chinese cultural perspectives regardless of the authors' membership. The books contained characters that were believable, fully developed, and show depth. The settings were natural in relation to the content and consistent with the time and place of Chinese/Chinese-American culture. Both narrative and dialogic language, as well as language in the glossaries, was accurate and natural. The themes were universal to all cultures and set accurately in China and in Chinatowns within North America. All cultural details were accurate, rich, and treated in depth, which helped the readers gain a sense of Chinese/Chinese-American culture. The illustrations presented accurate, natural, and non-stereotypical images of Chinese/Chinese-American people. The books were free from stereotype in any of the areas examined.

The 10 books in the middle of Table 9.3 were considered moderately authentic due to lack of depth in characters, setting, theme, cultural details, or unnatural use of language. All 10 books presented accurate, natural, non-stereotypical images of Chinese people. However, seven of the ten books focused more on explaining cultural details than telling stories of Chinese/Chinese American in daily life. The characters were so superficial that these books read like informational texts about Chinese New Year, food, or language. *Ling & Ting: Not Exactly the Same* (Lin, 2010) and *Ling & Ting Share a Birthday* (Lin, 2013) were what Bishop (1993) categorized as generic multicultural children's books. The books featured two Chinese-American girls, but contained few specific details

TABLE 9.3   Cultural Authenticity in Picture Books of Realistic Fiction

Level of Authenticity	Title	Author	Illustrator	Year
High	A New Year's Reunion	Li-Qiong Yu*	Zhu Cheng-Liang*	2011
	The Moon Lady	Amy Tan*	Gretchen Schields	1992
	Shanghai Messenger	Andrea Cheng	Ed Young*	2005
	Goldfish and Chrysanthemums	Andrea Cheng	Michelle Chang	2003
	Mei-Mei Loves the Morning	Margaret Tsubakiyama	Cornelius Van Wright & Ying-Hwa Hu*	1999
	Mama Bear	Chyng-Feng Sun*	Lolly Robinson	1994
	Henry's First-Moon Birthday	Lenore Look*	Yumi Heo*	2001
	Uncle Peter's Amazing Chinese Wedding	Lenore Look*	Yumi Heo*	2006
	Rabbit Mooncakes	Hoong Yee Lee Krakauer*		1994
	Sam and the Lucky Money	Karen Chinn	Cornelius Van Wright & Ying-Hwa Hu*	1995
	Our Home Is the Sea	Riki Levinson	Dennis Luzak	1988
	The Ugly Vegetables	Grace Lin*		1999
	Lion Dancer	Kate Waters & Madeline Slovenz-Low	Martha Cooper (photographer)	1990
Moderate	Chinatown	William Low*		1997
	Moy Moy	Leo Politi		1960
	This Next New Year	Janet Wong*	Yangsook Choi	2000
	Dim Sum For Everyone!	Grace Lin*		2001
	Kite Flying	Grace Lin*		2002

TABLE 9.3 Continued

Level of Authenticity	Title	Author	Illustrator	Year
	Bringing in the New Year	Grace Lin*		2008
	Ling & Ting: Not Exactly the Same	Grace Lin*		2010
	Ling & Ting Share a Birthday	Grace Lin*		2013
	In the Snow	Huy Voun Lee		1995
	Daisy Comes Home	Jan Brett		2002
Low	Tiger	Judy Allen	Tudor Humphries	1992
	Happy Belly, Happy Smile	Rachel Isadora		2009
	Hannah Is My Name	Belle Yang*		2004
	I Hate English!	Ellen Levine	Steve Björkman	1989
	Apple Pie 4th July	Janet S Wong*	Margret Chodos-Irvine	2002
	Chin Chiang and the Dragon's Dance	Ian Wallace		1984
	Lin Yi's Lantern: A Moon Festival Tale	Brenda Williams	Benjamin Lacombe	2009

*Chinese or Chinese American.

that were sufficient enough to define them as Chinese except their Asian-like images, name, and the food they ate. The last book, *Daisy Comes Home* (Brett, 2002), was a bit different from other books in this group. The author did a wonderful job in developing characters, creating specific settings in China through text and illustration. However, Daisy does not sound a real or authentic name for a hen in a small village in China.

The last seven books of realistic fiction showed a low level of authenticity because of the non-Chinese perspective reflected, presence of stereotypes, or inaccuracies in content and illustrations. *Tiger* (Allen & Humphries, 1992) was a story about a young Chinese boy who did not want people in his village to kill a tiger in the woods. The story was well crafted and realistically illustrated. However, it was written from the perspective of English-speaking culture because the conflict was built around the different meaning of the English word *shoot*. In English we can use the same word for two different meanings as in *shoot* the tiger with a camera and shoot the tiger with a gun. In Chinese, however, the two meanings have to be expressed by two different words instead of a homophone. Therefore, the story would have no suspense at all if it really happened in China as is set in the book. The second book, *Happy Belly, Happy Smile* (Isadora, 2009), is a story of a Chinese-American boy, who visits his grandfather's restaurant in Chinatown every Friday. The story reflects a very American perspective, particularly when indicating that Louie's favorite Chinese food is *egg rolls* and *shrimp chow mein*, food only served in American-Chinese restaurants. It is highly questionable that a Chinese-American child who has stayed close to Chinese culture would have the same preference. The next two books in this group are about two new immigrant Chinese girls. *Hannah Is My Name* (Yang, 2004) told the story of how a girl, who came to the United States with her mom and dad from Taiwan, had to give up her Chinese name and begun to use an American name, Hannah. In spite of the good intention to share an immigration story, the story reflected a very strong view of assimilation that in order to become American, immigrants had to give up their original identities. In addition, the description of the struggle of Hannah's family might leave the reader a false impression that only immigrants with green card could legally work in the United States, and hence perpetuated the existing stereotypes. *I Hate English* (Levine & Björkman, 1989) told the story of

Meimei, a new immigrant girl from Hong Kong who did not like English. The major problem with this book was the stereotype that new immigrants hated to learn English. In reality, many Chinese-American children are reluctant to speak Chinese, not English. Parents concerned about the loss of their culture heritage often force their children to learn Chinese by sending them to after-school or weekend Chinese schools (Cai, 2003). Another problem with the book was that the main character's conflict was resolved only when intervened by white characters, as noticed by Harada (1995). Consequently, reading the story perpetuates the stereotypes about Chinese immigrant children instead of promoting cultural understanding.

Three of the last seven books showed many inaccuracies in such areas as cultural details and illustrations. *Apple Pie 4th July* (Wong & Chodos-Irvine, 2002) was about a Chinese-American girl who was concerned that people would not buy the food her parents cooked on July 4th. The story seemed to have based on the assumption that people who ran Chinese restaurants would cook the food, and wait for people to come and buy, which was not true. In most dine-in and take-away Chinese restaurants, food is almost always cooked upon customers' order. Buffet restaurant or diners located in airports or food courts do have food readily cooked. However, there would be a wide variety of food to choose, instead of just one or two items. Another flaw of the story was, like *I Hate English,* the girl and her family did not solve their own problem, but rather relied on their white customers. *Chin Chiang and the Dragon's Dance* (Wallace, 1984) is another story about celebrating Chinese New Year. The story confused the dragon dance with the lion dance. Unlike the two people performing together in the story, Chinese dragon dance requires teamwork performed by many people, each of whom holds part of the dragon with a pole, with about five-to-six feet in between. The book also seemed to link dragon dance to the Year of Dragon. In reality, the dragon dance is a celebration traditionally performed at every Chinese New Year. *Lin Yi's Lantern: A Moon Festival Tale* (Williams & Lacombe, 2009) is a more recently published title. Unfortunately, it reflected ignorance about Chinese history and confusion between what was Chinese and what was Japanese. The author's note about market and bicycle in China indicated that the story happened in modern China. The main character, Lin Yi, was portrayed as a young boy wearing contemporary Chinese clothes and riding a bicycle. However,

other characters were either portrayed as living in several different historical times in ancient China, or as Japanese people wearing Kimonos. There
were also inaccurate cultural details. For example, Lin Yi asked a trader
the price of rice by *pound*, a measurement of weight which has never been
used in China. In addition, bicycles were not made in China and did not
become a common transportation vehicle until after 1949.

A theme common to the books of realistic fiction in the study is that
authors and illustrators are able to represent Chinese/Chinese-American
culture authentically or unauthentically regardless of their cultural membership. The difference lies in how much they have experienced with Chinese
culture.Overall, Chinese authors/illustrators are more likely to represent
authentic works about Chinese culture than their Chinese-American coun-
terparts and non-Chinese authors/illustrators. Non-Chinese authors/illustra-
tors who represented Chinese/Chinese-American culture authentically have
had direct or indirect experience with the culture. For example, Andrea
Cheng studied Chinese at Cornell and is married to a Chinese American;
Margaret Tsubakiyama taught English and studied Chinese language and
cooking for 1 year in Beijing; Riki Levinson visited Hong Kong regularly;
and Dennis Luzak went to Hong Kong to get a first-hand impression of the
city in order to accurately portray the characters and setting in *Our Home Is
the Sea* (Levinson & Luzak, 1988).Chinese-American authors may produce
works that are more culturally genetic, while non-Chinese authors' works are
more likely to reflect Euro-American perspective, contain inaccurate facts or
illustration. In addition, authors/illustrators born in China are more likely to
provide information about their cultural backgrounds than are America-born
Chinese authors/illustrators. Moreover, Asian authors/illustrators who are not
Chinese do not explicitly indicate who they are either. It might be a marketing
strategy of the publishers to remove explicit indicators of the author/illustra-
tor's identity and leave the reader to guess from information available such
as the author/illustrator's name and/or photo.

Another theme immerged from the analysis is the limitation of settings, topics, and occupations in these realistic stories. Twenty-three of the
30 stories are set in the United States and Canada, mostly in Chinatowns.
One-third of the books are about the Chinese New Year and Moon
Festival. The occupations of most adult characters in the realistic books
are not identifiable. Those that can be identified are associated with restaurants and Chinese Kung Fu. Chinese/Chinese-American children living

outside Chinatowns in the United States, or whose parents are working in other professional fields such as education, business, or natural science, are not represented in these books. As a result, the real-life experiences of Chinese/Chinese American and their contribution to the advancement of the U.S. societies continue remaining invisible in children's literature. In addition, the under-representation of Chinese people in contemporary China may keep depriving young readers from developing an up-to-date view of Chinese culture.

HISTORICAL FICTION

Ten books in this study were historical fiction, featuring stories in ancient or modern China as well as in the United States. Seven of the ten stories were written by Chinese and Chinese-American authors, and three by authors who were not Chinese/Chinese American.

As indicated in Table 9.4, six books represented Chinese/Chinese-American culture highly authentically. All six books accurately portrayed

TABLE 9.4   Cultural Authenticity in Picture Books of Historical Fiction

	Title	Author	Illustrator	Year
High	*Red Kite, Blue Kite*	Ji-li Jiang*	Greg Ruth	2013
	Little Eagle	Jiang Hong Chen* Translated by Claudia Zoe Bedrick		2007
	Landed	Milly Lee*	Yansook Choi	2006
	Henry and the Kite Dragon	Bruce Edward Hall*	William Low*	2004
	Nim and the War Effort	Milly Lee*	Yangsook Choi	1997
	Roses Sign on New Snow	Paul Yee*	Harvey Chan*	1991
Moderate	*Bitter Dumplings*	Jeanne Lee		2002
	Brothers	Yin*	Chris Soentpiet	2006
	Beautiful Warrior	Emily Arnold McCully		1998
Low	*Sparrow Girl*	Sara Pennypacker	Yoko Tanaka	2009

*Chinese or Chinese American.

experience of Chinese/Chinese Americans during such historical time as early years in the United States, World War II, and the Cultural Revolution. The characters were fully developed with believable and balanced personalities. The settings were consistent with the corresponding historical time. The language was natural and free from biases. The illustrations reflected realistic images of Chinese/Chinese Americans with a variety of physical characteristics. All six books were written by Chinese/Chinese-American authors, and all contained notes that provide historical background and contexts for the stories.

Three books of historical fiction showed some minor flaws and, therefore, were considered moderately authentic. *Bitter Dumplings* (Lee, 2002) was a story set during the years when Zheng He, a Chinese general in Ming Dynasty, and his fleet sailed seven times across the Indian and Pacific Oceans (1405–1433). The main character was driven out of home by her brother and sister-in-law and was adopted by an old woman selling bitter gourd dumplings. The problem was resolved when she met and married a handsome soldier who ran away from the military. Although no inaccuracies were noticed in setting, language, or illustrations, the story sounded like a Chinese Disney story and was moderate in authenticity at most. *Brothers* (Yin, 2006) described the struggle of early Chinese immigrants when they managed to survive on the new land. Examination of the book showed excellent results except that the Chinese boy had to rely on his white friends to bring business to his family grocery store. The last book in the moderate-authentic group, *Beautiful Warrior* (McCully, 1998), contained an inaccurate cultural fact. The information that Jingyong, the main character, was "born in the Forbidden City" in "the reign of the last Ming Emperor," to the father who was not the emperor, was not inconsistent with Chinese culture. The Forbidden City was the imperial palace and residence for only the emperor and his family members. Other people, except maids, eunuchs, and guards, could only enter when allowed. That was why it was called "Forbidden City."If Jingyong was born in the Forbidden City, she had to be a princess.

One historical fiction book turned out to have low authenticity. *Sparrow Girl* (Pennypacker & Tanaka, 2009) is one of the few books describing life in modern China after the founding of People's Republic of China in 1949. Despite a relatively accurate description of the early movement of *Eliminating the Four Pests* (i.e., rats, house sparrows, flies, and mosquitoes) in 1950s, the characters in the story were not portrayed as realistic Chinese people. They looked like white people except for their skin color. Dialogues in the story were not natural or authentic, failing to make the characters sound like Chinese people at all.

The majority of historical stories in this study were written by Chinese/ Chinese-American authors. Almost all contained some types of background notes about the historical backgrounds in which the story was set. Many topics covered by these 10 books have been rarely addressed in children's literature representing Chinese/Chinese-American culture and are valuable multicultural pieces to add to class libraries.

FANTASY

Only two books in this study are fantasies, and both raise major issues of authenticity in content and illustration. In *Emma's Dragon Hunt* (Stock, 1984), dragons were believed to cause solar eclipses and earthquake. The dragon was portrayed as a blue whiskered animal with white dots and wings, definitely foreign to Chinese/Chinese-American culture. The author's note mentioned that she heard the legend about dragons when visiting Hong Kong. However, there was no evidence of scholarly research beyond such superficial interaction about Chinese culture. *Fish in the Air* (1948) is another title by Kurt Wiese, the author and illustrator of the *Five Chinese Brothers*. Illustrations in the story showed the same stereotypical images of Chinese people—all characters look alike with yellow-tinted skin, bucked teeth, and slanted eyes. Several Chinese words on the plaques of the shops were misspelled. Neither of these two books represented Chinese/Chinese-American culture authentically.

Overall, of the 87 books examined in this study, 43 (49%) represented Chinese culture highly authentically, 25 (29%) books showed moderate level of authenticity, and 19 (22%) books showed relatively low level of authenticity. Lack of authenticity was presented by dominant white American perspectives, stereotypes about Chinese people and culture, and inaccurate factual details. Similar problems were found with illustrations, in the form of stereotypical portrayals of Chinese people, exotic depiction of backgrounds, or the presence of clothes and hairstyles belonging to different historical times in China within one story. The absence of authors' and illustrators' cultural backgrounds was consistent across genres. Consistent with the literature (Cai, 1994; Lin, 1999; Liu, 1998; Liu, 1993; Yokota, 1993), the findings of this study suggest that both Chinese and non-Chinese authors can present Chinese culture from insider perspectives. Nonetheless, Chinese/Chinese-American authors are more likely to create children's books representing Chinese culture authentically (68%) than non-Chinese authors (35%). Furthermore, authors who are outside the Chinese community but presented insider view in their books are likely to have done research about Chinese culture or have had first-hand experience of living in the culture. In contrast, non-Chinese authors who present outsider perspectives in their books are likely to have had little to no direct or indirect experience with the culture.

CONCLUDING REMARKS

As this study was concluding, two killers escaped from a federal prison in upstate New York. To make fun of the police, they left a yellow sticky note with a smiley face that looked like the racist images in *The Five Chinese Brothers*. Later in the summer at a New York county fair, yellow T-shirts with exactly the same smiley face were sold. This is the negative power of literature—it can create a stereotype and pass it down generation after generation without being noticed. Chinese/Chinese-American children who have never learned about what it is meant to represent may laugh at the image together with their white American peers. Those who have been told the discriminated connotation of the image may be afraid of being identified with the culture, or too uncomfortable to stop such a racist practice.

To disrupt and stop a stereotype begins with awareness, which can be gained through exposure to quality multicultural children's books in the classroom. If teachers want to use multicultural children's books as "mirrors" and "windows" for all children, then we have the responsibility to put mirrors in front of them that generate authentic reflections of the people who are looking into or out of the frames, rather than ones with distorting images. The earlier our children are introduced to multicultural children's literature with authentic cultural representations, the more likely they are to develop positive cultural identities as well as understanding of people who are different from them. Only in this way is there hope that children will be able to be aware that racist remarks and images are serious injustice. They should not be used for fun. They should be stopped. If the purpose of using multicultural children's literature is to help develop healthy cultural identities and dispel biases and stereotypes, it is important to put books with authentic representations to the hands of children. Picture books with moderate and poor authenticity can only be used productively when teachers understand what is not authentic and why, and when they are equipped with the tools to help children develop critical literacy skills and multicultural awareness.

The goal of this study is to share a specific Chinese view about children's picture books representing Chinese/Chinese-American culture. It is also meant to offer a tool that will assist teachers like *Ms. Kingsley* selecting quality children's literature to use in their classrooms. Given the complexity of cultural authenticity, it is challenging to use one single measurement to assess children books representing diverse cultures. The most useful tip for teachers to select authentic children's picture books may be Bishop's (2003) recommendation for reading as many books written by members from the culture depicted as possible. As we listen to people tell their own stories and what they think of their story when being told or retold by people who are outside their cultural community, we will have better opportunities to tell whether a children's book is or is not authentic.

KEYWORDS

- **Chinese culture**
- **picture books**
- **authenticity**
- **fiction**
- **Chinese-American authors**
- **stereotypes**

REFERENCES

1. Bishop, R.S. (1993). Multicultural literature for children: Making informed choices. In V.J. Harris (Ed.), *Teaching multicultural literature in grades K-8* (pp 37–53). Norwood, MA: Christopher-Gordon.
2. Bishop, R.S. (2003). Reframing the debate about cultural authenticity. In D. Fox & K. Short (Eds.), *Stories matter: The complexity of cultural authenticity in children's literature* (pp 25–37). Urbana, IL: National Council of Teachers of English.
3. Brinson, S.A. (2012). Knowledge of multicultural literature among early childhood educators. *Multicultural Education, 19*(2), 30–33.
4. Cai, M. (1994). Images of Chinese and Chinese Americans mirrored in picture books. *Children's Literature in Education, 25*(3), 169–191.
5. Cai, M. (2003). Can we fly across cultural gaps on the wings of imagination? Ethnicity, experience, and cultural authenticity. In D. Fox & K. Short (Eds.), *Stories matter: The complexity of cultural authenticity in children's literature* (pp 167–181). Urbana, IL: National Council of Teachers of English.
6. California State Department of Education. (1998). *10 quick ways to analyze children's books for racism and sexism.* Retrieved from http://cmascanada.ca/wp-content/uploads/2011/11/article-10-ways-to-analyze-childrens-books-for-sexism-and-racism.pdf
7. Colby, S.A., & Lyon, A.F. (2004). Heightening awareness about the importance of using multicultural literature. *Multicultural Education, 11*(3), 24–28.
8. Cooperative Children's Book Center School of Education, University of Wisconsin-Madison. (2014). *Children's books by and about people of color published in the United States.* Retrieved from http://ccbc.education.wisc.edu/books/pcstats.asp
9. Dong, L. (2013). Once upon a time in Chinese America: Chinese American folklore in American picture books. *Amerasia Journal, 39*(2), 48–70.
10. EunYoung, Y., Fowler, L., Adkins, D., Kyung-Sun, K., & Davis, H.N. (2014). Evaluating cultural authenticity in multicultural picture books: A collaborative analysis for diversity education. *Library Quarterly, 84*(3), 324–347.
11. Forum on Family and Child Statistics. (2014). *America's children in brief: Key national indicators of well-being.* Retrieved from http://www.childstats.gov/americaschildren/glance.asp
12. Fountas, I.C., & Pinnell, G.S. (2001). *Guiding readers and writers: Teaching comprehension, genre, and content literacy.* Portsmouth, NH: Heinemann.
13. Fountas, I.C., & Pinnell, G.S. (2006). *Teaching for comprehending and fluency: Thinking, talking, and writing about reading, K-8.* Portsmouth, NH: Heinemann.
14. Gangi, J.M. (2008). The unbearable whiteness of literacy instruction: Realizing the implications of the proficient reader research. *Multicultural Review, 17*(1), 30–35.
15. Hancock, M.J. (2000). *A celebration of literature and response: Children, books, and teachers in K-8 classrooms.* Upper Saddle River, NY: Merrill.
16. Harada, V.H. (1995). Issues of ethnicity, authenticity, and quality in Asian-American picture books, 1983–93. *Journal of Youth Services in Libraries, 8*(2), 135–149.
17. Hearne, B. (1993a). Cite the source. *School Library Journal, 39*(7), 22–27.
18. Hearne, B. (1993b). Respect the source: Reducing cultural chaos in picture books: Part two. *School Library Journal, 39*(8), 33–37.

19. Higgins, J.J. (2002). Multicultural children's literature: Creating and applying an evaluation tool in response to the needs of urban educators. *New Horizons for Learning.* Retrieved from http://education.jhu.edu/PD/newhorizons/strategies/topics/multicultural-education/multicultural-childrens-literature/

20. Howard, E.F. (1991). Authentic multicultural literature for children: An author's perspective. In M.V. Lindgren (Ed.), *The multicolored mirror: Cultural substance in literature for children and young adults* (pp 91–99). Fort Atkinson, WI: Highsmith.

21. Hsieh, I., & Matoush, M. (2012). Filial daughter, woman warrior, or identity-seeking fairytale princess: Fostering critical awareness through Mulan. *Children's Literature in Education, 43*(3), 213–222. doi:10.1007/s10583-011-9147-y.

22. IRA Notable Books for a Global Society Committee. (n.d.). *Outstanding K-12 multicultural literature: Criteria.* Retrieved from http://web.csulb.edu/org/childrens-lit/proj/nbgs/intro-nbgs.html#criteria

23. Klefstad, J.K., & Martinez, K.M. (2013). Promoting young children's cultural awareness and appreciation through multicultural books. *Young Children, 68*(5), 74–81.

24. Lasky, K. (2003). To stingo with love: An author's perspective on writing outside one's culture. In D. Fox & K. Short (Eds.), *Stories matter: The complexity of cultural authenticity in children's literature* (pp 84–92). Urbana, IL: National Council of Teachers of English.

25. Lin, J-S. (1961). *The milky way and other Chinese folk tales.* New York, NY: Harcourt, Brace & World.

26. Lin, Y. (1999). *Becoming illustrious: A study of illustrated Chinese American children's literature featuring female protagonists from 1963 to 1997* (Doctoral dissertation). Retrieved from ProQuest Dissertations & Theses Full Text. (9956290)

27. Liu, L. (1998). *Images of Chinese people, Chinese Americans, and Chinese culture in children's and adolescents' fiction (1980–1997)* (Doctoral dissertation). Retrieved from ProQuest Dissertations & Theses Full Text. (9841891)

28. Liu, M-Y. (1993). *Portrayals of the Chinese in fiction for children, 1925–1991* (Doctoral dissertation). Retrieved from ProQuest Dissertations & Theses Full Text. (9401307)

29. Loh-Hagan, V. v. (2014). A good year for Asian-American children's literature. *California Reader, 47*(3), 40–45.

30. Louie, B.Y. (2006). Guiding principles for teaching multicultural literature. *The Reading Teacher, 59*(5), 438–448. doi:10.1598/RT.59.5.3.

31. Mo, W., & Shen, W. (2003). Accuracy is not enough: The role of cultural values in the authenticity of picture books. In D. Fox & K. Short (Eds.), *Stories matter: The complexity of cultural authenticity in children's literature* (pp 198–212). Urbana, IL: National Council of Teachers of English.

32. Nilsson, N.L. (2005). How does Hispanic portrayal in children's books measure up after 40 years? The answer is "It depends." *The Reading Teacher, 58*(6), 534–548.

33. Noll, E. (2003). Accuracy and authenticity in American Indian children's literature. In D. Fox & K. Short (Eds.), *Stories matter: The complexity of cultural authenticity in children's literature* (pp 182–197). Urbana, IL: National Council of Teachers of English.

34. Norton, D.E. (2009). *Multicultural children's literature.* Upper Saddle River, NJ: Pearson.

35. Perini, R.L. (2002). The pearl in the shell: Author's notes in multicultural children literature. *The Reading Teacher, 55*(5), 428–432.
36. Rosenblatt, L.M. (1995). *Literature as exploration* (5th ed.). Chicago, IL: Modern Language Association.
37. Short, K.G., & Fox, D. (2003). The complexity of cultural authenticity in children's literature: Why the debates really matter. In D. Fox & K. Short (Eds.), *Stories matter: The complexity of cultural authenticity in children's literature* (pp 3–24). Urbana, IL: National Council of Teachers of English.
38. Smith, J., & Wiese, P. (2006). Authenticating children's literature: Raising cultural awareness with an inquiry-based project in a teacher education course. *Teacher Education Quarterly, 33*(2), 69–87.
39. Suh, B.K., & Samuel, F.A. (2011). The value of multiculturalism in a global village: In the context of teaching children's literature. *New England Reading Association Journal, 47*(1), 1–10.
40. U.S. Census Bureau. (2010). *The Asian population: 2010.* Retrieved from http://www.census.gov/prod/cen2010/briefs/c2010br-11.pdf
41. Wilfong, L.G. (2007). A mirror, a window: Assisting teachers in selecting appropriate multicultural young adult literature. *International Journal of Multicultural Education, 9*(1), 1–13.
42. Yokota, J. (1993). Issues in selecting multicultural children's literature. *Language Arts, 70*(3), 156–167.
43. Yokota, J., & Kolar, J. (2008). Advocating for peace and social justice through children's literature. *Social Studies & the Young Learner, 20*(3), 22–26.

Children's Picture Books Selected for Evaluation in the Study

Folktales

1. Bishop, C.H., & Wiese, H. (1938). *The five Chinese brothers*. New York, NY: Coward – McCann.
2. Casanova, M., & Young, E. (2000). *The hunter: A Chinese folktale*. New York, NY: Atheneum Books for Young Reader.
3. Chang, M.S., Chang, R., & Johnson, D. (1997). *The beggar's magic: A Chinese tale*. New York, NY: M.K. McElderry Books.
4. Chang, M.S., & Hutton, W. (1994). *The cricket warrior*. New York, NY: Margaret K. McElderry Books.
5. Chen, J.H. (2006). *The magic horse of Han Gan* (C.Z. Bedrick, Trans.). New York, NY: Enchanted Lion Books.
6. Compestine, Y.C., & Chau, T. (2001). *The runaway rice cake*. New York, NY: Simon & Schuster Books for Young Readers.
7. Compestine, Y.C., & Jorisch, S. (2007). *The real story of stone soup*. New York, NY: Penguin Group.
8. Compestine, Y.C., & Xuan, Y. (2001). *The story of chopsticks*. New York, NY: Holiday House.
9. Davol, M.W., & Sabuda, R. (1996). *The paper dragon*. New York, NY: Athenuem Books for Young Readers.

10. Demi. (1980). *Liang and the magic paintbrush.* New York, NY: Holt, Rinehart, and Winston.
11. Demi. (1987). *Chenping and his magic axe.* New York, NY: Holt.
12. Demi. (1990). *The empty pot.* New York, NY: Holt.
13. Demi. (1990). *The magic boat.* New York, NY: Holt.
14. Demi. (1994). *The magic tapestry: A Chinese folktale.* New York, NY: Holt.
15. Demi. (1998). *The greatest treasure.* New York, NY: Scholastic.
16. Demi. (2004). *The greatest power.* New York, NY: Margaret K. McElderry.
17. Drummond, A. (1992). *The willow pattern story.* New York, NY: North-South Books.
18. Greaves, M., & Gooding, B. (1985). *Once there were no pandas: A Chinese legend.* London, United Kingdom: Methuen.
19. Heyer, M., & Heyer, M. (1986). *The weaving of a dream: A Chinese folktale.* New York, NY: Viking Kestrel.
20. Hillman, E., & Wallner, J. (1992). *Min-Yo and the moon dragon.* San Diego CA: Harcourt Brace Jovanovich.
21. Hodges, M., & Young, E. (1989). *The voice of the great bell.* Boston, MA: Little Brown.
22. Kimishima, H., & Wakana, K. (1968). *Ma Lien and the magic brush* (A. Tresselt, Trans.). New York, NY: Parents' Magazine.
23. Kimmel, E.A., & Xuan, Y. (1998). *Ten suns: A Chinese legend.* New York, NY: Holiday House.
24. Lawson, J., & Morin, P. (Illustrator). (1993). *The dragon's pearl.* New York, NY: Clarion.
25. Leaf, M., & Young, E. (1987). *Eyes of the dragon.* New York, NY: Lothrop, Lee & Shepard Books.
26. Lee, J.M. (1982). *Legend of the Milky Way.* New York, NY: Holt, Rinehart, and Winston.
27. Louie, A-L., & Young, E. (1982). *Yeh-Shen: A Cinderella story from China.* New York, NY: Philomel Books.
28. Mahy, M., Tseng, J., & Tseng, M. (1990). *The seven Chinese brothers.* New York, NY: Scholastic.
29. Mosel, A., & Lent, B. (1968). *Tikki Tikki Tembo.* New York, NY: Holt, Rinehart and Winston.
30. Rappaport, D., & Yang, M-y. (1995). *The long-haired girl: A Chinese legend.* New York, NY: Dial Books for Young Readers.
31. Tseng, G., Tseng, J., & Tseng, M. (1999). *White tiger, blue serpent.* New York, NY: Lothrop, Lee & Shepard.
32. Williams, J., & Mayer, M. (1976). *Everyone knows what a dragon looks like.* New York, NY: Four Winds.
33. Wolkstein, D., & Parker, R.A. (1983). *The magic wings: A tale from China.* New York, NY: Dutton.
34. Wolkstein, D., & Young, E. (1972). *8,000 stones: A Chinese folktale.* Garden City, NY: Doubleday.
35. Wolkstein, D., & Young, E. (1979). *White wave: A Chinese tale.* New York, NY: Crowell.
36. Yacowitz, C., & Chen, J-H. (Illustrator). (1992). *The jade stone.* New York, NY: Holiday House.

37. Yep, L., & Mak, K. (1997). *Dragon prince: A Chinese beauty & beast tale.* New York, NY: HarperCollins.
38. Yep, L., & Seltzer, I. (1993). *The man who tricked a ghost.* Mahwah, NJ: Bridgewater Books.
39. Yep, L., Tseng, J., & Tseng, M. (1994). *The boy who swallowed snakes.* New York, NY: Scholastic.
40. Yolen, J., Charlip, R., & Maraslis, D. (1977). *The seeing stick.* New York, NY: Crowell.
41. Yolen, J., & Young, E. (1967). *The emperor and the kite.* Cleveland, OH: World.
42. Young, E. (1989). *Lon Po Po: A Red-Riding Hood story from China.* New York, NY: Philomel Books.
43. Young, E. (1995). *Night visitors.* New York, NY: Philomel Books.
44. Young, E. (1998). *The lost horse: A Chinese folktale.* San Diego, CA: Silver Whistle/ Harcourt Brace.
45. Young, E., Ed. (2001). *Monkey King.* New York, NY: HarperCollins.

Realistic Fiction

1. Allen, J., & Humphries, T. (1992). *Tiger.* Cambridge, MA: Candlewick.
2. Brett, J. (2002). *Daisy comes home.* New York, NY: Putnam.
3. Cheng, A., & Chang, M. (Illustrator). (2003). *Goldfish and Chrysanthemums.* New York, NY: Lee & Low Books.
4. Cheng, A., & Young, E. (2005). *Shanghai messenger.* New York, NY: Lee & Low Books.
5. Chinn, K., Wright, C.V., & Hu, Y-H. (1995). *Sam and the lucky money.* New York, NY: Lee & Low Books.
6. Isadora, R. (2009). *Happy belly, happy smile.* Boston, MA: Harcourt Children's Books.
7. Krakauer, H.Y.L. (1994). *Rabbit mooncakes.* Boston, MA: Joy Street Books.
8. Lee, H.V. (1995). *In the snow.* New York, NY: Holt.
9. Levine, E., & Björkman, S. (1989). *I hate English!* New York, NY: Scholastic.
10. Levinson, R., & Luzak, D. (1988). *Our home is the sea.* New York, NY: Dutton.
11. Lin, G. (1999). *The ugly vegetables.* Watertown, MA: Talewinds/Charlesbridge.
12. Lin, G. (2001). *Dim sum for everyone!* New York, NY: Knopf.
13. Lin, G. (2002). *Kite flying.* New York, NY:Knopf.
14. Lin, G. (2008). *Bringing in the New Year.* New York, NY: Knopf.
15. Lin, G. (2010). *Ling & Ting: Not exactly the same.* New York, NY: Little Brown.
16. Lin, G. (2013). *Ling & Ting share a birthday.* New York, NY: Little Brown.
17. Look, L., & Heo, Y. (2001). *Henry's first-moon birthday.* New York, NY: Atheneum Books for Young Readers.
18. Look, L., & Heo, Y. (2006). *Uncle Peter's amazing Chinese wedding.* New York, NY: Atheneum Books for Young Readers.
19. Low, W. (1997). *Chinatown.* New York NY: Holt.
20. Politi, L. (1960). *Moy Moy.* New York, NY: Scribner.
21. Sun, C-F., & Robinson, L. (1994). *Mama bear.* Boston, MA: Houghton Mifflin.
22. Tan, A., & Schields, G. (1992). *The Moon Lady.* New York, NY: Macmillan.

23. Tsubakiyama, M., Wright, C.V., & Hu, Y-H. (1999). *Mei-Mei loves the morning.* Morton Grove, IL: Albert Whitman.
24. Wallace, I. (1984). *Chin Chiang and the dragon's dance.* New York, NY: Atheneum Books for Young Reader.
25. Waters, K., Slovenz-Low, M., & Cooper, M. (1990). *Lion dancer: Ernie Wan's Chinese New Year.* New York, NY: Scholastic.
26. Williams, B., & Lacombe, B. (2009). *Lin Yi's lantern: A Moon Festival tale.* Cambridge, MA: Barefoot Books.
27. Wong, J.S., & Chodos-Irvine, M. (2002). *Apple pie 4th of July.* San Diego, CA: Harcourt.
28. Wong, J.S., & Choi, Y. (2000). *This next New Year.* New York, NY: Frances Foster Books.
29. Yang, B. (2004). *Hannah is my name.* Cambridge, MA: Candlewick.
30. Yu, L-Q., & Zhu, C-L. (2011). *A New Year's reunion.* Somerville, MA: Candlewick.

Historical Fiction

1. Chen, J.H. (2007). *Little Eagle* (C.Z. Bedrick, Trans.). New York, NY: Enchanted Lion Books.
2. Hall, B.E., & Low, W. (2004). *Henry and the kite dragon.* New York, NY: Philomel Books.
3. Jiang, J., & Ruth, G. (2013). *Red kite, blue kite.* New York, NY: Disney-Hyperion.
4. Lee, J.M. (2002). *Bitter dumplings.* New York, NY: Farrar, Straus and Giroux.
5. Lee, M., & Choi, Y. (1997). *Nim and the war effort.* New York, NY: Farrar, Straus and Giroux.
6. Lee, M., & Choi, Y. (2006). *Landed.* New York, NY: Farrar, Straus and Giroux.
7. McCully, E.A. (1998). *Beautiful warrior: The legend of the nun's kung fu.* New York, NY: Arthur A. Levine Books.
8. Pennypacker, S., & Tanaka, Y. (2009). *Sparrow girl.* New York, NY: Disney/Hyperion Books.
9. Yee, P., & Chan, H. (1991). *Roses sing on new snow: A delicious tale.* New York, NY: Maxwell Macmillan International.
10. Yin., & Soentpiet, C. (2006). *Brothers.* New York, NY: Philomel Books.

Fantasy

1. Stock, C. (1984). *Emma's dragon hunt.* New York, NY: Lothrop, Lee & Shepard.
2. Wiese, K. (1948). *Fish in the air.* New York, NY: Viking.

CHAPTER 10

POWER OF THE HEART AND LIMITATIONS OF LOVE: PRESERVICE TEACHERS' PERCEPTIONS OF THE TEACHING PROFESSION AND THEIR IMPLICATIONS

MAUREEN E. SQUIRES[1] and BRADLEY COUNTERMINE[2]

[1]*State University of New York at Plattsburgh, 101 Broad St, Plattsburgh, NY 12901, United States*

[2]*Beekmantown Middle School, 37 Eagle Way, West Chazy, NY 12992, United States*

ABSTRACT

Preservice teachers often enter the teaching profession because of preconceived notions of what it is to be a teacher. In this chapter, we explore the attitudes and conceptions of preservice teachers and their decision to become educators. Using data gleaned from qualitative surveys administered to students in a teacher preparation program that both did and did not disclose a disability, we trace the reasons why students entered the teaching preparation program and analyze their responses. Based on the survey results, we conceptualize an in-depth teacher education experience revolving around seminar-type courses and extended field experiences, while promoting the role of peer, faculty, and cooperating teacher mentors within the process. We believe addressing preconceived notions and turning theory into practice are the responsibilities of the teacher education program to prepare competent preservice teachers.

ACKNOWLEDGEMENTS

This research was funded by the Nuala McGann Drescher Award (New York State/United University Professions Affirmative Action Committee). Opinions expressed are solely those of the authors. The authors would like to thank Dr. Bev Burnell, Dr. Heidi Schnackenberg, and Ms. Cindy McCarty for their contributions to this study.

INTRODUCTION

Love is a powerful force. It is a factor necessary for teaching and learning, though often relegated to "the back burner," overshadowed by knowledge and skills. At a time when education is becoming increasingly standardized, data-driven, and focused on college and career readiness, there is a tendency to prioritize the cognitive domain at the expense of the affective domain. Quality teachers demonstrate mastery and integration of knowledge, skills, and positive dispositions. Hatt (2005) claims, "It is essential that a teacher be personally and professionally pre-disposed to loving children in their present circumstance and to loving the potential of becoming that resides within each of them" (p. 673). Yet, not any love will do. Pedagogical love extends beyond self-interest (which might be couched in good intentions) and provides the context for teachers and students to empower and transform each other. Pedagogical love is "'a creative, critical, and disruptive force' in education, possessing the capacity 'to fuel our intent to act against the barriers that block an abundant and engaged approach to teaching and learning'" (Liston & Garrison, as cited in Halpin, 2009, pp. 89–90).

This chapter focuses on the theme of love that emerged from a qualitative study, examining (1) the experiences of college students who self-identified with disabilities and (2) the perceptions of students without disabilities about working with differently abled individuals. The survey responses of 179 preservice teachers provide the foundation of this chapter. Two survey questions are particularly relevant. One question asked participants to describe the responsibilities of professionals in their field when working with individuals with disabilities. In addition to identifying skills and knowledge, many participants named essential dispositions, particularly qualities of care and compassion. Another question asked participants to describe how participants chose their major/profession. Participants indicated that they were influenced by family history, P-16 educational experiences, and personal reasons. Again, common affective variables (such as love for children and wanting to make a difference) were frequently identified by participants. Specifically, this chapter will (1) describe participants' responses to the aforementioned survey questions and (2) explore the ways participants use love to enact social justice for students with disabilities. This chapter will use critical theory to unpack meanings of love and include implications for teacher preparation programs.

LITERATURE REVIEW

RESPONSIBILITIES OF TEACHERS, PARTICULARLY DISPOSITIONS

The Council for the Accreditation of Educator Preparation (CAEP) recommends five standards by which quality teacher preparation programs are measured. The first standard, *content and pedagogical knowledge,* pertains to candidates' knowledge, skills, and dispositions regarding the learner and learning, content, instructional practice, and professional responsibility (CAEP, 2015). CAEP defines dispositions as "the habits of professional action and moral commitments that underlie an educator's performance" (CAEP, 2015, p. 122). The first CAEP standard also includes the 10 Interstate New Teacher Assessment and Support Consortium (InTASC) standards, which outline "the professional consensus [of] what new teachers should do and what they are capable of doing" (InTASC, 2015). In effect, the inTASC standards describe the responsibilities of teachers. Many of these responsibilities are rooted in the cognitive domain. New teachers must understand and apply their understandings of content areas, student development, diverse learners, evidence-based instructional strategies, management and motivation, technology and communication, instructional planning, instructional assessment, reflective practice, and school-community involvement. Lacking, at least on the surface, are responsibilities housed in the affective domain, often referred to as dispositions. Dispositions are an essential component of effective teaching. It is not enough to develop knowledge and skills. As Wise asserts (as cited in Mueller & Hindin, 2011), "Teacher preparation programs also must focus on the moral and ethical responsibilities of preservice teachers" (p. 17).

NCATE provides more clarity here, defining dispositions as the "professional attitudes, values, and beliefs demonstrated through both verbal and non-verbal behaviors as educators interact with students, families, colleagues, and communities ... which support student learning and development" (NCATE, 2008, pp. 89–90). NCATE (2008) stipulates that institutions "assess professional dispositions based on observable behaviors" (p. 90). Yet even this is limited. While attention to behavior is important, it cannot be the only focus. Dispositions are a combination of internal and external factors. Therefore, beliefs, attitudes, and actions must be examined and aligned in ways that positively affect learning outcomes of all students.

While experts do not agree on a single definition, they do agree on two points. First, dispositions are essential elements of quality education. In discussing teacher effectiveness, Taylor and Wasicsko (2000) focus on seven studies that indicate a relationship between teacher dispositions and student-learning outcomes. Throughout these studies, an ethic of care, compassion, and empathy frequently surfaced. Other qualities of effective teachers included using a democratic approach and believing that all people are worthy and deserve respect. Additional qualities, such as building and maintaining positive collaborative relationships, reflection, and continuous learning, often emerged. In their review of literature, Cummins and Asempapa (2013) arrive at the conclusion that "all researchers appeared to agree that dispositions were critical for teachers to success in the profession. Without effective dispositions, teaching and learning would suffer" (p. 102). Schussler and Knarr (2013) reach a similar conclusion. Based on their analysis of extant literature, they assert, "dispositions are an essential component of quality teaching and therefore must be cultivated to increase teacher effectiveness" (p. 72).

Second, researchers and theorists agree that simply defining or measuring dispositions is not enough for effective teaching and learning. Schussler and Knarr (2013) caution against restricting the construct of dispositions to a list of prescribed-desirable behaviors. This "reductionist approach disregards the intended purpose and limits the capacity of the construct to address more meaningful aspects of teaching, like the impetus driving one's behaviors and the purposes that person desires to achieve" (p. 72). They underscore the importance of self-knowledge of examining the underlying values and beliefs (the *why*) that contribute to teachers' actions (the *what*). Schussler and Knarr (2013) contend that with such self-awareness, teachers can first see and then address disconnects between intention, perception, and practice. Eliminating such gaps allows teachers to provide educational experiences that enable all students to learn. Schussler (2006), Cummins and Asempapa (2013), and Mueller and Hindin (2011) voice similar concerns. They call on teacher education programs to serve a greater purpose beyond compliance to external accountability standards or accreditation stipulations to fostering dispositions in ways that enhance teacher candidates' social consciousness and move them beyond rhetoric to actualizing inclusive, diverse, democratic educational spaces.

To foster positive dispositions in teacher candidates, institutions must take preservice teachers where they are and move them forward. After all, college freshmen do not arrive on campus as blank slates. They have already experienced typically 18 years of formal and informal education. As such, they enter college classrooms with preprogrammed, though not unchangeable, attitudes and inclinations. These dispositions are present when students make decisions about which majors to select (which professions to enter) and as they progress toward degree completion.

REASONS WHY PEOPLE ENTER THE TEACHING PROFESSION

People enter the teaching profession for a range of reasons. Decisions for becoming a teacher have both a practical and an emotional base. Lortie's (1975) seminal study (comprises a questionnaire survey and interview approach) on the attraction of teaching illustrates this point well. Lortie (1975) found five "attractors of teaching": the interpersonal theme (including relationships with youth, the love of teaching, and the love of children or adolescents), the service theme (where teachers are viewed as performing a mission or acting as stewards), the continuation theme (those who are attracted to teaching because their own P-12 experiences were positive and want to perpetuate the experience), the theme of time compatibility (work schedules that are flexible and family-friendly), and the material benefits theme (such as money, prestige, and job security).

Almost forty years later, Curtis's (2012) study involving approximately 1500 randomly selected secondary school mathematics teachers across the United States found similar affective characteristics. For instance, most teachers entered the profession "because of their desire to work with young people, love of mathematics, and reasons of personal fulfillment or making a difference" (p. 786). These teachers indicated (in descending order of frequency) that role model/teacher influence, lifestyle, contractual factors (e.g., benefits and working conditions), and parent–family encouragement influenced their decision to become an educator.

Mee, Haverback, and Passe (2012), using a case study approach, investigate why college students preparing to teach core subjects chose to enroll in the middle school major at a U.S. university. All participants reported selecting their major because they love the subject they plan to teach, have a calling to be a teacher, believe they can connect with young adolescents and make a difference in their lives, think there is a need to fill unpopular grade levels, and are influenced by friends and family members, regardless of whether these people are in the teaching profession (Mee, Haverback, & Passe, 2012).

These reasons why people select the teaching profession remain relatively consistent across demographic variables, such as age, gender, race, ethnicity, preferred teaching level, and content area. Taken together, these studies indicate that individuals choose to enter the teaching profession for a variety of reasons, including intrinsic (even altruistic) reasons and extrinsic reasons. There is no single factor that persuades a person to become a teacher. Yet, a common feature of the aforementioned studies is the presence of affective qualities in the decision-making process. Additionally, certain factors are especially relevant to P-16 education, which makes studying this topic important for practicing teachers, who may serve as role models for future teachers, and faculty in teacher preparation programs, who work closely with teacher candidates.

PHILOSOPHICAL AND THEORETICAL UNDERPINNINGS

We, the researchers, believe that all students can learn. We hold that teaching and learning should be individualized to meet students' developmental needs and that an asset-based model should be used in education. We deem that historically marginalized people, which include individuals with disabilities, deserve to have a voice. We also believe that education systems can always be renewed and improved.

Our philosophy has influenced the theoretical underpinnings of this study. As constructivists, we assume that human beings construct meanings of the world based on their perceptions. As such, knowledge is subjective and does not constitute a single absolute "truth" (Glesne, 2006). As critical theorists, we assume that research is situated in an environment influenced by myriad demographic factors. We also assume that society has perpetuated systems that privilege some human beings while disadvantaging others (Bogdan & Biklen, 1998; Glesne, 2006). We espouse critical research as an ethical and transformative act, one that McLaren has asserted can "empower the powerless" (as cited in Bogdan & Biklen, 1998, p. 21).

Accordingly, we designed a qualitative study using a phenomenological approach. This study is naturalistic; it includes descriptive empirical data, collected from interviews and open-ended surveys, attuned to the emic perspective; and it utilizes inductive methods where codes, categories, and themes emerge from data to construct meaning (Bogdan & Biklen, 1998; Creswell, 2012; Stake, 2010). As with qualitative research, a recursive process, data were analyzed multiple times with myriad frameworks.

METHODS: PARTICIPANTS, DATA COLLECTION, AND DATA ANALYSIS

The data analyzed for this chapter is part of a larger dataset (composed of 567 students enrolled in one of eight academic-professional preparation programs in a Northeastern U.S. comprehensive college). Participants (n=179) were purposefully selected so as to provide information-rich responses regarding the experiences and perceptions of preservice teachers (Glesne, 2006). Participants included males and females; undergraduate and graduate students; childhood, special, and adolescence education majors, with and without disabilities.

For this project, data collection and analysis involved a three-part open-ended survey. Part A contained demographic information from all participants, Part B was completed by students who disclosed having a disability, and Part C was completed by students without disabilities. Two questions were common to both Part B and Part C, with the majority of questions tailored to each group of respondents. Part B contained questions like "Have you disclosed that you have a disability to any individuals or offices on campus? Why or why not?" "What accommodations or supports have you or do you currently receive at this institution?" "Describe the steps you took to get these accommodations or supports." Part C contained question such as "Prior to entering your professional preparation program, what experiences have you had interacting with individuals with disabilities?" and "Describe the responsibilities of professionals in your field when treating or working with individuals with disabilities."

Demographic data from surveys were entered into a database; open-ended survey responses were typed, verbatim, into a separate database. All entered data were double checked for accuracy and then analyzed by emergent coding. Particular survey questions were identified for subsequent analysis. To establish inter-rater reliability, investigators independently read, coded, and categorized surveys and then came together to discuss their analyses. We decided which codes to adopt, eliminate, or revise and then defined codes/categories, thereby creating a coding scheme. All surveys were coded by at least two researchers using the same coding scheme.

FINDINGS

DECIDING ON A MAJOR IN EDUCATION

In the education field, the term "theory into practice" represents the shift from learning concrete, basic, and advanced skills about a particular field to internalizing generalized knowledge and putting it into practice on a daily basis. Teacher education programs attempt to bridge the gap between theory and practice through various field experiences, focusing on the theoretical underpinnings of education from objectives to content, educational aims to their corresponding results. To understand why teacher education students—those that did not disclose disabilities and those that did—came to decide on their professional preparation programs, the same short answer question, "How did you make the decision to choose your major?," was posed to teacher education students.

Of the 179 responses from teacher education students that *did not disclose* a disability, 12 respondents left the question blank and seven other respondents focused on logistical reasons for the choice, including "wanting to settle down," a "career change," and "limited choices." Adjusting for these responses, 160 responses remained for analysis from the "did not disclose" category. Of the 21 responses from teacher education students that *did disclose* a disability, every student responded to the question; however, only 19 responses were considered as two responses did not fit any of the codes. In total, 181 responses were analyzed from both the *did not disclose* and *did disclose* a disability category.

Respondents across the *did not disclose* and *did disclose* categories showed remarkable congruence in many responses. Both categories of respondents pointed to their experiences in P-12 schooling as a formative process whereby they determined they wanted to enter the teaching profession. Furthermore, both categories of respondents cited familial influence, work experience, and love of learning as an influence on their future professional choices. For the purposes of this study, the following codes were generated through open and axial coding. Some respondents completed their survey questions with nuanced answers, leading to multiple codes per response. Categories include *Family History* or how respondents viewed their family history as a genesis for their move toward the teaching profession, *School (P-16)* and its influence on respondents' choice to become professional educators, and *Personal* reasons for respondents to enter the teaching field. Each category contains subcategories to further illustrate the nuanced responses gleaned from the respondents. For the purposes of this chapter, we focused on *Personal* reasons. Presentation and analysis of the findings allow an opportunity to understand why students with and without disabilities entered the teacher preparation program.

EXPERIENCE WITH INDIVIDUALS WITH DISABILITIES

Within responses from both categories, respondents pointed to both positive and negative experiences with individuals with disabilities. Though respondents mentioned individuals with disabilities in their justification for entering the profession, family members with disabilities (or in the case of three respondents from the *did disclose* category, alluding to their own disability as a reason to enter the profession and help students with disabilities) did not have a large influence on entering the teaching profession. *Did not disclose* respondents cited work experience (both volunteer and school-based) far more often than family members with disabilities as their motivation to enter the teaching profession.

POSITIVE EXPERIENCES

Responses from both categories focused on positive experiences interacting with family members with disabilities. A key difference between the two categories was respondents in the *did not disclose* category focused on a family member and his/her disability as a positive event in shaping their decision to enter teaching. One respondent from the *did not disclose category* responded, "How much early intervention really helped support my brother's foundational skills," and another response explained, "My uncle with Down Syndrome was my inspiration."

Respondents in the *did disclose* category used their disability as a reason to enter the teaching profession. All three respondents had valuable reasons for entering the profession. One respondent answered, "When finding out about my ADHD, I knew this would just help me advocate for my students." Another respondent wrote, "I've always wanted to become a teacher and would love to work with students that have special needs because I know where they are coming from since I myself have a disability." The third respondent that pointed to his/her disability stated, "I have a disability so I know firsthand how a student might feel and I can help them." Each of the three respondents from the *did disclose* category that pointed to their disability as a reason to become teachers all framed their disability as a positive. They would "understand" and "advocate" for students with disabilities because of their lived experience with a disability.

NEGATIVE EXPERIENCES

No students in the *did disclose* category wrote of negative family experiences as a reason for entering the profession. Two respondents in the *did not disclose* category illustrated negative family experiences. Both respondents centered on the experience their cousin's had in relation to their disabilities. One respondent wrote, "...seeing the struggles my cousin faces encouraged me to try to make changes." The other respondent replied, "My cousin who was diagnosed with a mental disability and cerebral palsy was underestimated and unsupported during his school time. I want to help prevent these situations." Both respondents characterized their cousins' negative interaction with the larger world, one generally with "struggles my cousin faces" and another specifically with "[he] was underestimated and unsupported during his school time." The respondents used these experiences to explain their choice to enter the field and remarked that they "wanted to try to make changes," and "want to help prevent these [negative] situations." Remarkably, these responses only accounted for 1% of total responses for entering the profession, though respondents from both categories frequently emphasized "making a difference" as their reason to pursue a career in education (see section "Personal Reasons").

TEACHER INFLUENCE

Respondents in both categories mentioned the effect their former teachers had on their choice to become educators. Both participants in the *did disclose* category that credited teachers believed they shaped their future. One respondent wrote, "My teachers were my role models," another responded, "I liked my content teachers in high school." In the *did not disclose* category, 14 participants agreed that teachers made the difference in their career choice. Comments from these respondents included, "[My] science teacher in middle school really made a connection and helped me get through his subject," "I had a lot of teachers who inspired me," "I was having a difficult time in high school and my biggest support system came from my teachers and I want to give that opportunity to someone else," "My teachers in middle school and high school were my role models and I wanted to impact other people's lives like they've impacted mine," and "I had many role model teachers in elementary school." Whether it was elementary school, middle school, or high school, these 14 respondents believed that teachers and the attention they gave them during the tumultuous times of growing up made an impact on their career choices.

ENJOYED SCHOOL

Respondents from both categories did not factor their school enjoyment into their decision to become educators. In the *did disclose* category, only one respondent discussed a love of school as the reason he/she decided to be a teacher. Three respondents in the *did not disclose* category determined that enjoying school was an underlying reason why they began the teacher education program.

PERSONAL REASONS

Overwhelmingly, respondents from both the *did disclose* and *did not disclose* categories cited personal reasons for entering the teaching profession. Within *personal reasons*, codes were generated for *previous work experiences, always wanted to teach*, and *to make a difference*. Among the three sub-themes, preservice teachers in both categories cited "making a difference" more often than "always wanting to teach" and "previous work experiences," though preservice teachers in the *did disclose* category discussed "always wanting to teach" seven times and "making a difference" six times for a total of 68% of all responses. Respondents from the *did not disclose* category also considered "always wanting to teach" and "to make a difference" as important factors, with 40 "always wanting to teach" and 43 "making a difference." Overall, 52% of respondents cited *Personal Reasons*. Respondents from the *did not disclose* category differed from their peers in the did disclose category in that 39 responses or 24% attributed their professional aspirations to *previous work experiences*, while respondents that *did disclose* only explained *previous work experiences* four times (11%) in their responses.

PREVIOUS WORK EXPERIENCES

Previous work experiences encapsulated a variety of responses from the *did disclose* and *did not disclose* participants. Examples of responses from the *did disclose* category include nonrelated teaching jobs like "Working with teenagers" and "I volunteered at a daycare and elementary schools." Participants in the *did not disclose* category had a wide variety of responses. Specifically, respondents identified volunteer work done in special education classrooms, "I did an internship in high school and there was a student with Autism in the classroom. He was the student who made me want to go into the special education field," working at summer camps, "teaching dance to those with disabilities," and "babysitting" as reasons why they wanted to prepare for the teaching profession.

Whether respondents first worked with children as babysitters, camp counselors, or preschool room workers, they developed a relationship with working with students that translated to career paths in education. Multiple respondents (five) in the *did not disclose* category linked their time volunteering in classrooms to their chosen career. Furthermore, participants in both categories often pointed to prior work experience (both formal and informal) in their decision to become educators.

ALWAYS WANTED TO TEACH

Respondents in both categories reported, "always wanting to teach." Seven respondents in the *did disclose* category spoke of "passion," "being fulfilled," and maintained, "I always knew I wanted to teach." In the *did not disclose* category, 40 respondents spoke of the same feelings, "passion," "love," and, "This is something I've always dreamed of doing."

Survey participants from both categories felt a calling to the educational profession. They viewed the profession not as a job, but as their duty to students full of "love" and "passion." According to the respondents, teaching as a career was something they knew they wanted to do long before they entered college.

TO MAKE A DIFFERENCE

The only response more prevalent than "always wanting to teach" was "to make a difference." Six respondents in the *did disclose* category wanted to make a difference in their future students' lives. Two particularly strike quotes read, "I realized I wanted two things: to stay as close to literature and writing and to make an impact on the world" and "My life will not be fulfilled until I have changed a life for the better."

Forty-three respondents in the *did not disclose* category came to the teacher education program in an effort to "make a difference." Sample responses include, "I want to help and be a role model to others. [I want] to make a difference in someone's life," "I love teaching and have always wanted to impact others," "I enjoy being around children and I want a job that I feel like I am helping someone and making a difference," and "I've always had a wish to change someone's life for the better, and I feel like I could absolutely do that in this profession." Of the 43 respondents, six specifically mentioned "making a difference," while the others wanted to "make an impact" and "always want[ing] to help people."

SUMMARY

Respondents in both categories pointed to various reasons for wanting to enter the teaching profession. Though survey answers ranged from *family history* responses to *school* themes, a majority of respondents identified personal themes as the reason they sought to become teachers. Of the *personal* themes, respondents identified "making a difference" and "always wanting to teach" as significant reasons for choosing their major.

The potentially problematic language of "helping" others places these respondents in a perceived position of power. Many of their responses read as if they want to "save" the students and "make a difference" in their lives. While outwardly appealing as a reason to become teachers, the notion of "saving" students and "making a difference" denotes an educational system that needs fixing. If preservice teachers enter the profession with plans to revamp, revise, and reconstitute the teaching profession, the level of cognitive dissonance between their theoretical approach to the profession, "to save students" and "make a difference" and the actuality of their place within the system as novice teachers in an increasingly deskilled environment could lead to burnout and disengagement from the very profession they profess "passion" and "love" for.

RESPONSIBILITIES OF TEACHERS

Teaching is a social activity, consisting of interactions among multiple people on multiple levels. An ecological framework suggests that human activity occurs through these multiple contexts, which in turn exert influence on the human being. As Bronfenbrenner (2005) writes, "Human development occurs in the midst of a vibrant, complex environment... [an environment] largely defined by social and cultural practices and institutions that provide most of the experiences that people have" (p. 3). Teacher candidates are no different. Their perceptions of the responsibilities of effective teachers touch upon multiple systems, including themselves, their students, parents/guardians, service providers, and the community/public. As such, their role as advocate encompasses numerous elements.

One-hundred-seventy-nine participants in teacher education completed survey question #7 (*Describe the responsibilities of professionals in your field when working with individuals with disabilities*). Of these, 31 participants left this question blank and five participants indicated they were unsure or did not know, yielding 142 surveys for analysis. Participants' responses ranged in specificity; some responses were multifaceted and thereby received multiple codes.

Three-hundred-eighty-eight comments were made about the knowledge, skills, and dispositions of teachers who work with students with disabilities. *Knowledge* refers to pedagogical knowledge, professional knowledge, and student knowledge. *Skill* refers to performance, the ability to apply information in a particular context. *Dispositions* are the professional attitudes, values, and beliefs displayed through verbal and nonverbal interactions. What follows is a presentation and analysis of findings, organized by the three aforementioned categories.

SKILLS

Participants identified numerous skills teachers who work with individuals with disabilities must demonstrate. These skills are captured in the following categories: "teaching," "building relationships," and "advocating."

TEACHING

Participants described this responsibility in several ways. In general terms, they wrote that teachers "deliver services" and "deliver instruction" to students. This model positions the teacher as dispenser of information and the student as receptacle in which knowledge is deposited. Participants described the teacher's goal as "to build students' living skills and social skills" or "to improve students' content knowledge and related skills." Some participants elaborated on particular skill sets (e.g., cooking, cleaning, and laundry). Education in this broad sense was mentioned 21 times. Participants explained that "helping students grow" is achieved by "working with students' IEPs and annual goals" and "adapting lesson plans and the classroom atmosphere." They wrote about individualized learning plans and differentiating instruction 37 times throughout surveys. These were the most popular emergent themes regarding the act of teaching. In addition, five participants referred to "tracking student progress" and "recording data" to ensure that instruction was positively related to student learning. This continues to illustrate the unidirectional influence of the teacher on the leaner. In opposition to this model, five participants wrote about the teacher's responsibility as "informing" and "educating" the student about his/her disability. In this way, the student is gaining important self-knowledge that he/she can use to inform his/her decisions about his/her own life. Self-awareness can influence self-regulation, which enhances one's independence.

While the majority of participants' responses centered on teaching the student with the disability, their comments extended to other individuals, as well. In particular, they wrote about educating peers, service providers, parents/guardians, and the local community. Participants elaborated on the education of parents/guardians, explaining that teachers should make parents/guardians aware of their child's disability, regularly share information about their child's performance and progress, provide information on relevant resources and organizations, and ensure that parents/guardians understand their rights (e.g., informed consent and participation) and the rights of their child. One participant summarized this by stating, "Knowledge is power and sometimes families are not even aware of the resources available … Teachers should be able to provide that knowledge to parents."

BUILDING RELATIONSHIPS

Another skill frequently mentioned (20 times) by participants is building relationships with students, parents/guardians, and colleagues. Participants described effective relationships as "good," "close," or "strong." They elaborated: positive relationships require "open" and "clear" lines of communication. One participant even stated that "Communication and honesty with everyone (family, teachers, therapist) is extremely important … If one doesn't help, it hinders everything." Additionally, participants explained that effective communication was multidirectional. Teachers could "communicate with the guardian for information"; teachers could "meet with parents to give updates"; and teachers could "collaborate with special educators, school psychologists and family members to help the student." Other service providers mentioned include school administrators, physical therapists, and occupational therapists.

Student relationships, in particular, are grounded in *respect* and *patience* (each mentioned 14 times in survey responses). Respect meant "listening" to students and honoring their needs and requests. One participant summarized this by writing, "Students with disabilities are important. Their voices should be heard." It also means being "professional" and "maintaining confidentiality." *Patience* requires the teacher to see the value in each student, to "pay full attention," "to allow the individual time without rushing him/her," and "to treat students with disabilities how they want to be treated." Such comments illustrate teachers who are resolved to help students grow toward independence and follow student-determined aspirations at a pace and on a path appropriate for each student. By demonstrating respect and patience, teachers were supporting their students.

ADVOCATING

Participants commented about advocacy 61 times in surveys, using the words "advocate," "advocating," or "advocacy." Some participants applied advocacy to pedagogy. For instance, one described his/her responsibility as "advocating for a child's education/learning needs." Similarly, another wrote that being an advocate meant "providing the best possible care." Another described advocacy as "helping students receive the support they need." Other participants saw advocacy as a means to equality. For example, one explained that teachers should "make sure that students with disabilities are given the same opportunities as other students." Another wrote about the importance of "making sure students with disabilities receive a just education." Yet another discussed "advocating for change." Common to the later meanings of advocacy is the notion of social justice.

There was little variation in what participants meant by advocacy. All but once, participants identified the teacher as the advocate. One participant shared, "Professionals in this field should advocate for these children [students with disabilities] as if the child were their own." Another claimed, "It is the teacher's job to not only speak for all individuals with disabilities but also specific students." In this sense, the teacher is presented as guardian and savior of the student with a disability. Only once did a participant mention the notion of students with disabilities as "self-advocates." Additionally, participants mentioned advocating for parents and their rights (seven times). They also explained that advocacy needed to occur within "the public" and "the community." According to participants, this would require teachers to educate society about disabilities and to teach others acceptable ways of speaking about and interacting with individuals with disabilities.

DISPOSITIONS

Numerous dispositions were discussed throughout surveys. The following categories represent themes relevant to love and compassion.

BEING HELPFUL AND WANTING THE BEST FOR STUDENTS

Participants often commented that a responsibility of teachers was "to be helpful." Participants typically not only directed help toward the student but also used the term when speaking about families and education service providers. Participants also noted that teachers should want "the best" for their students. They applied this concept in many ways. Participants referred to what they expected of themselves: providing the best educational opportunities, individualized plans, care, and learning environment. They also noted what was expected of students: to work to "the best of their ability." In this way, both teachers and students brought their "A-game" to the classroom. Being helpful and wanting good things for fellow human beings require a disposition toward altruism and a belief in human agency.

POSSESSING COMPASSION

In the category of *possessing compassion*, participants wrote about "being understanding," "caring for students," and "having love for children." Of these comments, *being understanding* was most often identified (by nine participants). This meant understanding that students might have struggles, supporting students, and helping students when necessary, which require empathy. *Caring for* individuals was mentioned by five participants. This concept was used in a nurturing way to illustrate the loving manner a teacher has for his/her student. For instance, one participant wrote, "The responsibilities for people in this profession to people with disabilities or not is to be loving, caring." It also was used in a tending way to exemplify the stance a teacher takes when helping a student in need. For example, another participant wrote, "Taking care and making sure the student receives everything they need." This includes appropriate services both at school and at home. *Having love for children* was mentioned by four participants. This includes "compassion," "kindness," and "gentleness." Notions of compassion surfaced repeatedly throughout surveys, sometimes multiple times by a single participant. They were important based on the number of times mentioned and the length and detail of response provided.

BEING PROTECTIVE

Another complementary theme is that of *being protective*. Beyond making students feel safe and welcome in the classroom, other participants described the teacher as a guardian, one who "looks out for" students with disabilities. A cursory glance at this image suggests a caring relationship between the teacher and student, one where the teacher champions the student. A critical examination, however, reveals a limiting depiction of students with disabilities. It positions individuals with disabilities as weak, in need of defense of a stronger more powerful person. In this case, teachers serve as their shield.

DISCUSSION/IMPLICATIONS

Throughout the responses ran an altruistic undercurrent. Whether the participants came to teaching to make a difference or advocate for others, they made the decision consciously to help others. However, within the helping perspective and their approach to teaching, these participants brought preconceived notions of teaching that, if uncorrected, could lead to unintended consequences.

WHAT LOVE ISN'T

Though preservice teachers came to the teaching profession to "make a difference" and because "they always wanted to," their ideas of "making a difference" because they "always wanted to" are fraught with the contradictions of putting theory into practice, particularly within the shift from student in an education program to novice teacher. For example, in the responses, many students spoke of their role as an "advocate" for their future students. Though teachers should advocate for their students from a caring perspective, without teaching self-regulation and self-advocacy skills, students do not become independent, they become *more* dependent on their "advocate," reinforcing the feeling of "making a difference" at the expense of actual actualized independent growth.

Furthermore, "love" in this sense, while a seemingly innocuous perspective that all teachers should have toward their students, becomes a potential enabler, stunting student growth and providing opportunities for the teacher to "fix" the student and "become happy." By tying self-worth to student success, preservice teachers set their students and themselves up for potential failure.

Indeed, responses, such as "Because teaching is the only thing that will make me happy" and "My life will not be fulfilled until I have changed a life for the better," focus on the *teacher* rather than on the student. Disassociating from this perspective celebrates hidden achievements of students not necessarily sought after within the mainstream educational standards movement (e.g., Bloom's Affective Domain dispositions, positive self-talk, positive self-concept, self-regulation, and coping strategies necessary for success in the world after school).

To "love" students means to provide for the gradual release of responsibility from the teacher to the student, building self-regulation skills and student confidence in the process. An essential element of pedagogical love is the relationship between the teacher and the student. Cho (2005), adopting Valenzuela's conceptual framework, claimed, "The proper pedagogy of care is the one that passes from aesthetic to authentic care" (p. 87). This is where teachers shift their focus from the technical duties of the job (teaching, grading, correcting, behavior, etc.) to the reciprocal relationships between themselves and their students. It is a particular type of caring relationship that fosters the creative and expansive development of children and adolescents—and does so in a humanizing way, without the potentially patronizing effects of a skewed perspective based on the teacher's perspective of the special education student as "needing help."

Freire (2006) vehemently criticized the teacher–student dichotomy. He declared, "Education must begin with the solution of the teacher-student contradiction, by reconciling the poles of the contradiction so that both are simultaneously teachers and students" (p. 72). In this reciprocal relationship, teachers *and* students work in solidarity. Freire (2006) explains, "Here, no one teaches another, nor is anyone self-taught. People teach each other, mediated by the world" (p. 80). As such, teachers are not presumed omnipotent; and students are not presumed ignorant beings to be filled with knowledge like buckets to be filled with water. The images of teacher as active-subject and student as passive-object do not exist. Power does not reside in the teacher; and the student is not subordinate. Such relationships, Freire (2006) maintains, are authentic and liberatory; they are essential in education. It is particularly important that teachers remain cognizant of the reality that students with special needs do not need to be saved and taken care of, they need to develop skills to approach their world successfully in conjunction with the teacher as a guide and a facilitator, not as a savior.

SHAPING TEACHER EDUCATION

Considering that many preservice teachers identify "making a difference" as their reason for entering the profession, teacher education programs must respond to be responsive of student needs. Teaching from a critical perspective, basing coursework on both the theoretical underpinnings of teaching and learning and the practical aspects of the job, allows for preservice teachers to confront the reality of a career in education. Following are recommendations for a reflexive and responsive teacher education program.

As Freire (2006) and others (e.g., Apple, Darder, Giroux, Hooks, McLaren) explain, critical discourse, through generative questions and in-depth discussions of events, feelings, and actions, allows for growth. Structuring seminar courses throughout the teacher education program to address preservice teachers' preconceived notions and perceptions can create critical growth. Within the seminar courses, topics can be discussed in a logical progression beginning with the Foundations of Education followed by the Theories of Education, the study of the Practice of Education, and finally Guided Reflexive Practice during the formal student teaching experience. Using Bruner's (1960) concept of the spiraled curriculum, preservice teachers focus on specific aspects of teaching and learning throughout their learning experience.

Potential topics taught in the Foundations of Education seminar could revolve around the historical need for formal education, the history of education in the United States, the function of education from various perspectives, and the development of a personal teaching and learning philosophy. In an associated field experience, students could not only look for examples of these topics but they could also develop questions for consideration during the seminar sessions where ideas can be refined and applied to the Theories of Education seminar course. Within the first field experience, students would be paired in twos and placed in a best-practices classroom with a mentor teacher, college mentoring in the form of the seminar experience, combined with peer support in the partner groupings.

In the Theories of Education seminar, students would continue to develop their perspective of teaching and learning. Revisiting topics on the function of education and adding child and adolescent psychological development would create an environment that links teaching and learning to the wider world and provides a space for considering the effect of public schools on society and society's influence on public schools. The associated field experience would continue in the previous field experience classroom with the mentor teacher (based upon successful completion of the first field experience) and focus students on the "big picture" ideas behind schooling. Preservice teachers could study educational reform efforts and observe their corresponding influence on public schools. Through interviews with teachers, principals, parents, and students, preservice teachers could determine how the foundations of education affect theories of education and begin to pair contemporary theories of education with the corresponding practice of education in the next seminar course.

The Practice of Education seminar continues the sequence begun by the Foundations of Education and Theories of Education course. Once students develop baseline skills and competencies, both knowledge-based and instruction-based, preservice teachers take an even more active role in their learning experience. During the Practice of Education course, preservice teachers determine who they are as a professional drawing upon their previous experiences in the field, with mentor teachers, professors, and peers. Topics in this course would include in-depth learning and practice on classroom management, lesson planning, differentiated instruction, technology usage in the classroom, and the holistic aspect of the teaching experience (e.g., teaching responsibilities, paperwork responsibilities, and service responsibilities). The field experience would draw upon these developed skills and, continuing the cohort model, would preferably include the same peers, mentor teachers, and professors as the previous courses.

The capstone course for the teacher education program would be the Reflective/Reflexive Practice Seminar course, the student teaching experience. In this seminar course, students would be placed with their peers and mentor teachers to implement the knowledge honed throughout the previous seminar courses. Seminar sessions would focus on experience in the field and collaboration between preservice teachers to tackle general and specific issues within their placements. Drawing the previous seminars together, this culminating experience would prepare preservice teachers for the realities of the teaching profession.

CONCLUSION

The reconciliation [between the] … apparent opposition between the ideas of self-activity and external direction … lies in the fact that the educator supplies conditions in which the pupil responds, so that his activity is directed indirectly. The educator's problem here is simply to surround the child with those conditions or that environment which will arouse in him in due proportion the various elements in the natural organic activity (Dewey, 1902, para. 9).

To account for preservice teacher preconceived notions toward the profession, a fundamental shift must be undertaken to provide more experiential learning opportunities to complement theoretical knowledge developed through teacher education coursework. Using preservice teacher attitudes toward the profession as a starting point and providing them with learning opportunities to develop pedagogical love (Hoveid & Finne, 2014), preservice teachers can move from a "helping" and "making a difference" space to a more appropriate "facilitating" and "generating" perspective toward special education students.

It is the teacher education program's responsibility to shape the preservice teacher learning experience in a meaningful, structured, and relevant way. If we, as teacher educators, do not plan and develop practical experiences to challenge preservice teachers' tendencies toward patronizing perspectives, we will promote and create a profession that continues the status quo, with teachers "helping" students succeed. We must begin the process of promoting introspection, reflection, self-regulation, and self-esteem within the preservice teacher corps using their original link to teacher education. However, we should shift their perspective away from ideas of "saving" students and move toward empowering preservice teachers to be the difference without adversely affecting special education student development. Providing authentic teaching and learning experiences allows for preservice teacher transcendence above the rhetoric of "making a difference" to actually changing their future students' lives.

KEYWORDS

- **Dispositions**
- **Knowledge**
- **Love**
- **Preservice Teachers**
- **Skills**
- **Special Education**
- **Teacher Education**

REFERENCES

1. Bogdan, R., & Biklen, S. (1998). *Qualitative research for education: An introduction to theory and methods* (3rd ed.). Boston: Allyn & Bacon.
2. Bronfenbrenner, U. (2005). Ecological models of human development. In M. Gauvain & M. Cole (Eds.), *Reading on the development of children* (4th ed., pp. 3–8). New York: Worth Publishers.
3. Bruner, J. (1960). *The process of education.* Cambridge, MA: The President and Fellows of Harvard College.
4. Cho, D. (2005). Lessons of love: Psychoanalysis and teacher-student love. *Educational Theory, 55*(1), 79–95.
5. Council for the Accreditation of Educator Preparation (CAEP). (2015). *CAEP Accreditation Manual (Draft Version 2).* Washington, DC: Author. Retrieved from http://caepnet.org/~/media/Files/caep/accreditation-resources/caep-accreditation-manual.pdf?la=en
6. Creswell, J.W. (2012). *Qualitative Inquiry & Research Design* (3rd ed.). Thousand Oaks, CA: SAGE Publications, Inc.
7. Cummins, L., & Asempapa, B. (2013). Fostering teacher candidate dispositions in teacher education programs. *Journal of Scholarship of Teaching and Learning, 13*(3), 99–119.
8. Curtis, C. (2012). Why do they choose to teach – and why do they leave? A study of middle school and high school mathematics teachers. *Education, 132*(4), 779–788.
9. Dewey, J. (1902, June 9). Principles of education: The psychology of growth, continued. In L. A. Hickman (Ed.), (in press), *Class lecture notes of John Dewey: Volume 2: Education, logic, and social and political philosophy: The electronic edition.* Charlottesville, VA: InteLex Corp.
10. Freire, P. (2006). *Pedagogy of the oppressed* (30th anniversary edition). New York: Continuum.
11. Glesne, C. (2006). *Becoming qualitative researchers: An introduction* (3rd ed.). New York: Pearson.
12. Halpin, D. (2009). Pedagogy of *Romantic* love. *Pedagogy, Culture, & Society, 17*(1), pp. 89–102.
13. Hatt, B.E. (2005). Pedagogical love in the transactional curriculum. *Journal of Curriculum Studies, 37*(6), 671–688.
14. Hoveid, M.H., & Finne, A. (2014). 'You have to give of yourself': Care and love in pedagogical relations. *Journal of Philosophy of Education, 48*(2), 247–259.
15. Interstate New Teacher Assessment and Support Consortium (InTASC). (2015). *About InTASC Standards.* Retrieved from http://intascstandards.net/about-intasc-standards/.
16. Lortie, D.C. (1975). *Schoolteacher.* Chicago: The University of Chicago Press.
17. Mee, M., Haverback, H.R., & Passe, J. (2012). For the love of the middle: A glimpse into why one group of preservice teachers chose middle grades education. *Middle Grades Research Journal, 7*(4), 1–14.
18. Mueller, M., & Hindin, A. (2011). An analysis of the factors that influence preservice elementary teachers' developing dispositions about teaching all children. *Issues in Teacher Education, 20*(1), 17–34.

19. National Council for Accreditation of Teacher Education (NCATE). (2008). *Professional Standards for the Accreditation of Teacher Preparation Institutions*. Washington, DC: Author. Retrieved from http://ncate.org/LinkClick.aspx?fileticket=nX43fwKc4Ak%3d &tabid=474

20. Schussler, D.L. (2006). Defining dispositions: Wading through murky waters. *The Teacher Educator*, 41(4), 251–268.

21. Schussler, D.L., & Knarr, L. (2013). Building awareness of dispositions: Enhancing moral sensibilities in teaching. *Journal of Moral Education*, 42(1), 71–87.

22. Stake, R.E. (2010). *Qualitative research: studying how things work*. New York: Guilford Press.

23. Taylor, R.L., & Wasicsko, M.M. (2000). The dispositions to teach. Retrieved from https://coehs.nku.edu/content/dam/coehs/docs/dispositions/resources/The_Dispositons_to_Teach.pdf

INDEX

Printed in the United States
by Baker & Taylor Publisher Services